COLLINS
GEM

BASIC FACTS

CHEMISTRY

W A H Scott BSc PhD
Adviser
A Brookes BSc PhD

Collins
London and Glasgow

First published 1982
Reprint 10 9 8 7 6 5

© Wm. Collins Sons & Co. Ltd. 1982

ISBN 0 00 458887 8

Phototypeset and illustrated by
Parkway Illustrated Press

Printed in Great Britain by Collins, Glasgow

Introduction

Basic Facts is a new generation of illustrated GEM dictionaries in important school subjects. They cover all the important ideas and topics in these subjects up to the level of first examinations.

Bold words in an entry means a word or idea developed further in a separate entry: *italic* words are highlighted for importance.

Elements are presented in this consistent format:

The A$_r$ value of the element

The Atomic number of the element

The element's density (g/cm^3)

The symbol of the element

The element's melting point

The element's boiling point

electron configuration

Series adviser: K. Davison B.Sc., M.Sc., Dip. Ed., M.I. Biol.

A$_r$ This is the symbol for relative atomic mass. The A$_r$ value for an **element** is a measure of the mass of an atom of the element compared to the mass of an atom of the **isotope** carbon-12. See **M$_r$** and **Atomic mass**.

Absolute temperature See **Kelvin temperature scale**.

Abundance The measure of how much of a substance exists. It is often expressed as a percentage.
Elements The abundances of **elements** in the earth and in the sun are shown in the table.

Earth elements	O	Si	Al	Fe	Ca	Na	K	Mg
% abundance	46.6	27.7	8.1	5.0	3.6	2.8	2.6	2.1

Sun elements	H	He	O	C	Si
% abundance	54.0	44.7	0.8	0.4	0.05

Isotopes The abundances of **isotopes** in some commonly occurring elements are shown in the table.

element	% abundance of each isotope	
Chlorine	$^{35}_{17}$Cl [75.5%]	$^{37}_{17}$Cl [24.5%]
Carbon	$^{12}_{6}$C [98.9%]	$^{13}_{6}$C [1.1%]
Magnesium	$^{24}_{12}$Mg [78.6%]	$^{25}_{12}$Mg [10.1%]
		$^{26}_{12}$Mg [11.3%]

Knowing the percentage abundance of each isotope it is possible to calculate the A_r value for the element.

Accumulator This is a rechargeable **battery**. The most common type is the lead-acid accumulator. This is the battery which is used to start cars and lorrys. It is also used to drive milk floats and stacker trucks. In the future we shall probably drive electric cars which are powered by accumulators.

In the lead-acid accumulator the **electrolyte** is dilute **sulphuric acid**. The positive plate is made of lead(IV) oxide and the negative plate is composed of lead. When the accumulator is discharged (flat) the plates are coated with lead(II) sulphate. The sulphate coating is removed when the accumulator is recharged by passing electricity through it. The lead-acid accumulator is dangerous because it can produce very high currents. It produces the flammable gas **hydrogen** whilst it is being recharged and there is the constant danger of the **acid** spilling from the accumulator.

The nickel-cadmium accumulator is 'dry'. The positive plate is made of a complex nickel salt (nickel oxyhydroxide), the negative plate is made up of cadmium. The electrolyte is **potassium hydroxide** solution which is soaked into a spongy material and so cannot spill. Nickel-cadmium accumulators work well at low temperatures but, as yet, are very expensive.

Acid An acid is a substance which releases **hydrogen ions** (H^+) when it is added to **water**. The hydrogen ion is solvated to give the **oxonium ion** (H_3O^+). Acidic solutions have a **pH** of less than 7.

Common laboratory acids are:
   Nitric acid HNO_3
   Hydrochloric acid HCl
   Sulphuric acid H_2SO_4
   Ethanoic acid CH_3COOH

These acids are dangerous, corrosive liquids and should always be treated with care.

Acids:

★ turn *blue* **litmus** *red*

★ give **carbon dioxide** when added to **carbonates**

★ give **hydrogen** when added to certain **metals**

★ neutralize **alkalis**

Acid-base reactions Acids react with *bases* to form a *salt* and water only. Some examples are shown here:

Hydrochloric acid HCl(aq)	+ Sodium hydroxide NaOH(aq)	→ Sodium chloride NaCl(aq)	+ water $H_2O(l)$

Nitric acid $2HNO_3(aq)$	+ Copper(II) oxide CuO(s)	→ Copper(II) nitrate $Cu(NO_3)_2(aq)$	+ water $H_2O(l)$

Acid salts These are **salts** in which only some of the **hydrogen atoms** in an **acid molecule** have been replaced by a metal. Some examples are shown below:

normal salt		acid salt	
Na_2CO_3	Sodium carbonate	$NaHCO_3$	Sodium hydrogencarbonate
Na_2SO_4	Sodium sulphate	$NaHSO_4$	Sodium hydrogensulphate
Na_3PO_4	Sodium phosphate	Na_2HPO_4	Sodium hydrogenphosphate

4

Acidic oxides Most **oxides** of **non-metals** react with water to form acidic **solutions**. Because of this they are known as acidic oxides.
Examples

oxide		acid	
Carbon dioxide	CO_2	Carbonic acid	H_2CO_3
Sulphur dioxide	SO_2	Sulphurous acid	H_2SO_3
Sulphur trioxide	SO_3	Sulphuric acid	H_2SO_4

See **Basic oxide** and **Neutral** oxide.

Activation energy When **hydrogen** and **oxygen** are mixed there is no **reaction**. When a **flame** is brought into contact with the gases, there is an immediate **explosion**. The heat of the flame causes the reaction to occur. Many chemical reactions are like this. It seems that there is a barrier which has to be overcome before the reaction will take place. **Energy** (e.g. the heat of the flame) has to be put in to make the reaction occur. This energy is known as the activation energy. See **Catalyst**.

Addition reaction   These are reactions in which two or more molecules are reacted together to form a single molecule. Good examples are found in the reactions of **alkenes**. These compounds contain **double bonds** and can add an atom to each side of the double bond to form a **saturated alkane**. The example of **ethene** reacting with **hydrogen** is shown here.

ethene + hydrogen⟶ ethane

Addition reactions also occur in the making of many **polymers**, e.g. **poly(ethene)**.

Air   This is the name given to the **mixture** of gases which surrounds the earth. The average composition of pure air is shown on the next page. This composition varies from place to place and also varies with altitude.

oxygen (20.9%)
argon (0.9%)
carbon dioxide (0.03%)
nitrgoen (78.1%)

Other components

neon
helium
krypton
xenon
} In small constant amounts.

water In very variable amounts.

Air also contains pollutants, some of which are shown below.

pollutant	sources
Sulphur dioxide	Burning coal, oil.
Carbon monoxide	Engines, cigarettes.
Oxides of nitrogen	Car engines.
Soot	Fires, engines.
Pollen	Trees, flowers.
Dust	Volcanoes.

Air is vital for life. The **oxygen** is necessary for **respiration** and the **carbon dioxide** needed for **photosynthesis**.

Alcohols Alcohols are important **organic** compounds. They possess the arrangement of atoms

$$\diagdown \atop {-C-O}^{-H} \diagup$$

Ethanol is the most important compound but see also **methanol** and **glycol**.

Alcohols with a small M_r value are flammable

liquids which dissolve in water. Some of the important properties of alcohols are that they can be oxidized easily. They form **esters** and react with **sodium** to produce **hydrogen**.

Alkali   An alkali is a base which is soluble in water. They are usually metal **hydroxides**, e.g. **sodium hydroxide**, but **ammonia** solution is also an alkali. Their reactions are affected by their *strength* and their *concentration*.

Oven cleaners, household **ammonia** and some paint strippers contain alkalis.

Alkalis:

★ turn red **litmus** blue

★ neutralize **acids**

★ have a **pH** above 7

★ react with acids to produce a **salt** and **water** only

Alkanes   Alkanes are **hydrocarbons**. They are very important compounds in our lives. Our gas supplies (North Sea gas and 'Calor' gas) are almost 100% alkanes. Vaseline or petroleum jelly is made of alkanes. The chief source of alkanes is **petroleum** and they are valuable raw materials in the chemical industry.

They have a general formula: C_nH_{2n+2} so the alkane with 4 carbon atoms will have 10 hydrogen atoms: C_4H_{10}, **butane**.

See also **Methane, Ethane** and **Propane**. They

are **saturated** compounds and so are not very reactive. They tend to be **flammable** and will react with **chlorine** in **ultra-violet light**.

Alkenes Alkenes are important **hydrocarbon** compounds. They are widely made in oil-refineries and are used as starting materials in the manufacture of many materials, e.g. **plastics**. **Ethene**, propene and styrene (phenylethene) are three of the most important alkenes.

They have a general formula: C_nH_{2n}. So the alkene with three carbon atoms (propene) will have six hydrogen atoms. Because they are **unsaturated** compounds they have a carbon-carbon **double bond** and are reactive compounds. Typically they will undergo **addition reactions** with hydrogen, **halogens** and water. They will also form **polymers** by this addition process.

See **Poly(phenylethene)** and **Poly(propene)**.

Alkynes There is only one common alkyne. This is **ethyne**. Alkynes have the general formula C_nH_{2n-2} and so, ethyne with 2 carbon atoms will have 2 hydrogen atoms: C_2H_2. Alkynes have carbon-carbon **triple bonds** and therefore show the typical **addition reactions** of **unsaturated** compounds.

Allotropes Some elements exist in different forms in the same physical state. The chemical properties are the same but the physical **properties** are different. The best example is carbon. There are two allotropes: **diamond** and **graphite**. Diamond is hard and colourless, graphite is flakey and black. Other elements which exist in different allotropic forms are **sulphur**, **phosphorus** and **tin**.

Alloy   An alloy is a **mixture** which is made up of two or more **metals** or which contains metals and **non-metals**.

Alloys are much more widely used than pure metals are, because by bringing two or more elements together in the right proportions specific properties can be obtained. For example, aluminium is quite a soft metal but when a small amount of copper is mixed in, the alloy duralumin is strong enough to be used in aircraft frames. Some common examples are: **brass**, **bronze**, **duralumin**, **pewter**, solder and **steel**.

Alpha particles $^4_2He^{2+}$ These are **helium** atoms without their electrons. They are produced in many **nuclear reactions**. They have a fairly short range in air (usually less than 10 cm) and are easily stopped by thin sheets of paper or foil.

Aluminium   Aluminium is the most abundant **metal** in the earth's crust (approximately 8 per cent by mass). Its **alloys** are widely used at home and in industry; the diagram below shows some of its uses.

Clay, shale, slate and granite all contain aluminium compounds but the metal is difficult to extract from them. It is obtained by the **electrolysis** of **bauxite** dissolved in **cryolite**.

The metal has valency 3. Although it is a very reactive metal, it, and its alloys, are resistant to **corrosion** because of a layer of **oxide** on the surface of the metal. It is subject to **anodizing**.

Aluminium compounds

Aluminium chloride $AlCl_3$	The **anhydrous** chloride is **covalent**.
Aluminium oxide Al_2O_3	The oxide is **amphoteric**.
Aluminium sulphate $Al_2(SO_4)_3$	This is the most important aluminium compound. It is used as a precipitator in sewage works, as a mordant and as a size in the paper industry. It is also used as a foaming agent in fire-extinguishers.

chocolate wrapping

Coca-cola can

saucepan

pressure pan

window frame

milk-bottle top

aluminium kitchen foil

christmas decorations

sardine tin

Some uses of aluminium

12

Amalgam An amalgam is an **alloy** which contains **mercury**. Zinc amalgam is used for teeth fillings. Amalgamation was once used to extract gold and silver from crushed rock or earth which contained the metals.

Amino acids These are **organic** compounds which contain $-COOH$ and $-NH_2$ groups. They are the building bricks from which **proteins** are made. Twenty different amino acids are needed to make up all the proteins in our bodies. Some of these amino acids our bodies can produce but there are eight which we need to eat – the *essential* amino acids. Amino acids combine together to form **peptides**.

glycine

Ammonia (NH_3)   Ammonia is a colourless gas with a pungent odour. It is very soluble in water giving an *alkaline* solution. Ammonia was once made from coal but now over 90% comes from the **Haber process**.

Ammonia is also produced by bacteria found on the roots of leguminous plants (peas and beans). When protein decomposes, ammonia is released. Both of these are important sources of plant food. See the **nitrogen cycle**.

The importance of ammonia in our lives is shown in the diagram below.

Fertilizers [80% of production]
(ammonium sulphate)
(ammonium phosphate)
(ammonium nitrate)

Pharmaceuticals

Cleanser in the Textile industry

Household cleanser

Ammonia

Manufacture of **Nylon**

Urea
(used to make plastics and resins)

Refrigerators

Explosives
(**Ammonium nitrate**)
(TNT)

Sodium Carbonate
(Solvay process)

Nitric acid

The uses of ammonia

14

Ammonia is a **covalent** compound. It has a characteristic reaction with **hydrogen chloride** to give dense white fumes of ammonium chloride:

$$NH_3(g) + HCl(g) \rightarrow NH_4Cl(s)$$

Ammonia can be oxidized to **nitric acid** and this is the major source of nitric acid today. It is easily liquefied and it is usually carried from place to place in tankers in the liquid form.

Ammonium compounds

Ammonium carbonate $(NH_4)_2CO_3$	An unstable compound used as smelling salts *sal volatile*.
Ammonium chloride NH_4Cl	A salt which undergoes **dissociation** into ammonia and hydrogen chloride.
Ammonium nitrate NH_4NO_3	A very important compound which is used as a fertilizer and an explosive.
Ammonium phosphate $(NH_4)_3PO_4$	A convenient chemical with which to put both nitrogen and phosphorus into the soil.
Ammonium sulphate $(NH_4)_2SO_4$	An important fertilizer.

Ammonium hydroxide (NH_4OH) This term was once used to describe a solution of **ammonia** in water. The term ammonia solution is now preferred.

Ammonium ion Ammonium compounds contain the ammonium ion as does a solution of ammonia gas in water. The ion is **positively** charged and the hydrogen atoms are arranged around the nitrogen atom in a **tetrahedral** way.

NH_4^+

Amphoteric (oxides and hydroxides) Insoluble **oxides** and **hydroxides** which show both basic and acidic **properties** are said to be amphoteric. The examples shown here will react with both **acids** and **alkalis** to form **salts**.

	Examples	
Al_2O_3	PbO	ZnO
$Al(OH)_3$	$Pb(OH)_2$	$Zn(OH)_2$

For example:
$$ZnO + 2HCl \rightarrow ZnCl_2 + 2H_2O$$
$$ZnO + 2NaOH + H_2O \rightarrow Na_2Zn(OH)_4$$
$$\text{(sodium zincate)}$$

The solid **hydroxides** of these metals all dissolve in **sodium hydroxide** solution.

Amorphous Material which is non-crystalline is described as amorphous; a-morphous = without definite shape or form.

Ampère (amp) This is the **unit** of electric current. It measures the rate of flow of charge. One amp = 1 **coulomb/second**.

Anhydrides These are substances which react with water to produce acids. All **acidic oxides** are anhydrides, but some will give rise to one acid only whilst others will produce two.

Anhydride	Examples	Acid
SO_3		H_2SO_4
NO_2		$\begin{cases} HNO_3 \\ HNO_2 \end{cases}$

Anhydrous Something which contains no water is said to be anhydrous. The word is usually used about salts which have no **water of crystallization**.

For example:

Anhydrous copper(II) sulphate $CuSO_4$

Anhydrous sodium carbonate Na_2CO_3

The **hydrated** salts are:

$$CuSO_4.\ 5H_2O$$
$$Na_2CO_3.\ 10H_2O$$

The term is also used about liquids which are perfectly dry, e.g. anhydrous ether.

17

Anion Anions are negatively charged **ions**. They usually contain a full outer orbital of electrons. Non metals form anions but complex ions can also be anions.

For example:
Sulphate SO_4^{2-} Manganate(VII) MnO_4^-
Carbonate OH^- Nitrate NO_3^-

Annealing Heating and cooling **metals** causes changes in their properties, especially their strength. When a metal is heated to a high temperature and then cooled *very* slowly the metal becomes soft and **malleable**. This is because the **crystals** in the metal have been allowed to grow very large. This slow-cooling is called annealing.

Anode When a solution undergoes **electrolysis** the **electrode** with the *positive* potential is called the anode. **Anions** are attracted to the anode because they are negatively charged. The ions give up their extra **electrons** at the anode. Non-metallic elements are produced and the electrons travel round the circuit. See **Cathode**.

Anodizing **Aluminium** has a protective layer of oxide (Al_2O_3). This layer of oxide can be thickened by making the piece of aluminium the **anode** (+) in an **electrolysis cell**. Dilute **sulphuric acid** is used as the **electrolyte**. Aluminium atoms give up electrons and they react with the water:

$$2Al(s) + 3H_2O(l) \rightarrow$$
$$Al_2O_3(s) + 6H^+(aq) + 6 \text{ electrons}^-.$$

Aluminium oxide is formed and this can be dyed to produce attractive finishes to products, e.g. coloured milk bottle tops.

Antifreeze This is a substance which is added to the cooling system of engines in winter to prevent the formation of **ice** which would damage the engine. The substances added *lower* the freezing point of the water. The two most commonly used substances are:

Methanol $CH_3 - OH$

Ethane $- 1, 2 -$ diol

$$\underset{CH_2{-}CH_2}{\overset{HO \qquad\qquad OH}{}}$$

aq This subscript is used to denote an **aqueous solution**, e.g. NaCl(aq).

Aqueous solution This is a solution where **water** is the **solvent**.

Argon (Ar) Argon is the most abundant **inert gas**. It makes up 0.9% of the **atmosphere** (by volume). It is very unreactive. No **compound** of argon has yet been made.

Argon has uses where the presence of an unreactive gas is useful, e.g. light bulbs and fluorescent tubes.

Aromatic This term is used to describe compounds which contain a **benzene** ring structure in their molecules.

Arsenic Arsenic is a brittle grey **metalloid**. Its compounds are very poisonous – especially the oxide (As_2O_3). It is a group v element which occurs as **allotropes**. Arsenic compounds are used in pesticides. The grey allotrope sublimes at $613\,°C$.

Asbestos This is a naturally occurring silicate which was widely used in the past for fire-proof clothing and insulating materials. Its use today is restricted because of the dangers of contracting *asbestosis*, a fatal lung disease. It was much used in schools as bench mats and centre pieces for gauzes. These are now made of material which does not contain asbestos.

Atmosphere 1) A unit of **pressure**. Although the pressure of the air which surrounds us varies from place to place and from time to time its value usually only varies slightly and is always approximately one atmosphere.

2) A term used to describe the mixture of gases which surrounds a planet, sun or moon. The atmosphere on earth is **air**.

Atom Atoms are the building bricks with which everything is made. Atoms are small particles; 100 million placed end-to-end would measure 1 cm. They are made up of even smaller – sub-atomic – particles:

Protons	These particles are found in the
Neutrons }	centre of the atom – the **nucleus**.
Electrons	These particles move round the
	nucleus.

The atom as a whole is electrically **neutral** although the protons and electrons carry electrical charges. Atoms which lose or gain electrons are called **ions**. All atoms of the same **element** contain the same number of protons and have the same **atomic number**, but atoms of the same element can contain different numbers of neutrons. See **Isotope**.

An atom can be thought of as the smallest part of an element that can take part in a chemical reaction.

Atomic mass Atoms have different masses if they contain different numbers of **protons, neutrons** and **electrons**. Because these particles are so small the mass of an atom is tiny. It is not usual to refer to the mass of an atom in grams or kilograms but we normally compare the mass of one atom with a standard mass. This is called the **relative atomic mass (A_r)**.

Atomic number This is the number of **protons** in the **nucleus** of an atom. All atoms of the same **element** have the same atomic number, e.g. sodium atoms all contain 11 protons.

In a **neutral** atom the number of electrons equals the atomic number.

Atomicity This is the number of **atoms** in a

molecule of an element. With the **inert gases** there is only one atom in the molecule and the atomicity is 1. All other gases have an atomicity of 2, e.g. **oxygen** (O_2), **nitrogen** (N_2), except **ozone** (O_3). Molecules in solid **sulphur** have an atomicity of 8 (S_8).

Avogadro constant/number (L)   This is the number of **atoms** that are contained in exactly 12 g of the carbon-12 **isotope**. More generally it is the number of particles present in a **mole** of substance. $L = 6 \times 10^{23} \, mol^{-1}$.

Avogadro's hypothesis/law   This states that when **temperature** and **pressure** are the same, equal **volumes** of gases contain the same number of **molecules**, e.g. 10 **dm³** of oxygen and 10 dm³ of hydrogen both contain the same number of molecules. They contain twice as many molecules as 5 dm³ of chlorine – provided that all the volumes were measured at the same temperature and pressure.

Baking powder   A mixture of a carbonate and a weak acid which is used in cooking. When water is added to the solid mixture or the mixture is heated, **carbon dioxide** is produced. This produces bubbles in the dough or cake mixture and it *rises*, producing a product with an open texture. The mixture usually contains sodium hydrogencarbonate ($NaHCO_3$) and **tartaric acid**.

23

The powder can be made up in the kitchen by mixing cream of tarter (the potassium salt of tartaric acid) and baking soda (sodium hydrogencarbonate). This prevents the reaction from occurring before it is required.

Balance 1) A device for comparing the masses of objects. It is usual to use electric balances in schools these days.

2) Chemical **equations** must be balanced. It is important to have equal numbers of each atom on each side of the equation as atoms can neither be created nor destroyed.

Example:

$$Zn(s) + HCl(aq) \rightarrow ZnCl_2(aq) + H_2(g)$$

No. of atoms:

$Zn = 1\ H = 1\ Cl = 1\ Zn = 1\ Cl = 2\ H = 2$

The equation *is not* balanced.

$$Zn(s) + 2HCl(aq) \rightarrow ZnCl_2(aq) + H_2(g)$$

The equation *is* balanced.

Barium   Barium is a very reactive **group II element**. It occurs in nature as the **sulphate** (barytes) and as the **carbonate**. The **metal** is extracted from the molten **chloride** by electrolysis.

Barium chloride solution is used to test for sulphates.

Barrel 1) A cask for holding beer. It has a **volume** of exactly 32 imperial gallons.

2) A unit of volume in the **petroleum** industry. One barrel = approximately thirty-five imperial gallons. (159 litres).

Base A base is a substance which reacts with an acid to form a **salt** and **water** only. Bases are usually **metal oxides** or **hydroxides**, e.g. sodium hydroxide (NaOH) and copper(II) oxide (CuO).

$$NaOH(aq) + HCl(aq) \rightarrow NaCl(aq) + H_2O(l)$$
$$CuO(s) + H_2SO_4(aq) \rightarrow CuSO_4(aq) + H_2O(l)$$

Metal oxides and hydroxides which are soluble in water are known as **alkalis**. These are compounds of **group** I and II metals.

Ammonia solution is an alkali. It contains hydroxide ions in *equilibrium*:

$$NH_3 + H_2O \rightleftharpoons NH_4 + OH^-$$

Basic oxide An oxide which reacts with an acid to form a **salt** and **water** only. Basic oxides are oxides of metals, but all metals *do not* give basic oxides. See **Amphoteric oxide** and **Acidic oxide**.

25

Basic salts These are **salts** which contain **hydroxide ions** as well as normal **anions**, e.g. **sulphate, carbonate**.

Common examples are:

Malachite	$CuCO_3.Cu(OH)_2$
Azurite	$2CuCO_3.Cu(OH)_2$
White lead	$2PbCO_3.Pb(OH)_2$

Battery A device which converts chemical **energy** into electrical energy. Chemical reactions occur and, in them, **electrons** are sent through a circuit. Batteries are of two types: the rechargeable and the non-rechargeable. The former kind are known as **accumulators**.

Batteries are available in different shapes, sizes and price ranges and their chemical composition varies too. The common 'dry battery' has **carbon** and **zinc electrodes** and a manganese(IV) oxide **electrolyte**. It produces a **voltage** of 1.5 volts. See **Cell**.

Bauxite Bauxite is the chief **ore** from which **aluminium** is extracted. It is a **hydrated** oxide (Al_2O_3. x H_2O) and is found in tropical regions of the world, e.g. Northern Australia and West Africa.

Bee-hive shelf This piece of apparatus is used to allow a gas to be collected over water.

trough with water

Benzene Benzene (C_6H_6) is the simplest **aromatic** compound. It is a toxic liquid **hydrocarbon** which gives rise to the production of cancers. Its widespread use in schools has been replaced by that of **methylbenzene**.

Beta Particles These are **electrons** which are produced in the following **nuclear reaction**:

$$^1_0n \rightarrow {}^1_1p + {}^{\,\,0}_{-1}e$$

neutron proton electron

In this radioactive decay the electrons are expelled from the **nucleus** of the atom. They travel at high speeds (up to 98% of the speed of light) and have a greater penetrating power than **alpha particles**. When this reaction occurs, the atom concerned changes its **atomic number** because it gains a **proton**.

Bitumen   This is a mixture of high boiling-point hydrocarbons which is left behind in the distillation of **petroleum**. It is used for roofing and road surfacing.

Blast furnace   The blast furnace allows a continuous production of molten iron. At the bottom of the furnace the coke and hot air react:

$$C(s) + O_2(g) \rightarrow CO_2(g)$$

This **exothermic** reaction raises the temperature to 1800°C. The coke then reacts with the carbon dioxide:

$$C(s) + CO_2(g) \rightarrow 2CO(g)$$

The carbon monoxide formed reduces the iron oxides to iron (1200°C). For example:

$$Fe_2O_3(s) + 3CO(g) \rightarrow 3CO_2(g) + 2Fe(s)$$

The molten iron flows to the bottom of the furnace and the limestone is broken down into calcium oxide and carbon dioxide.

The calcium oxide reacts with impurities from the iron ore (mainly silica (SiO_2)) and forms a molten slag:

$$CaO(s) + SiO_2(s) \rightarrow CaSiO_3(l)$$

This sinks to the bottom of the furnace where it floats on top of the iron. The iron which is produced contains about 3% **carbon** and the **metal** is brittle. In **steel** making the amount of impurities in the iron ore is very carefully controlled.

Bleach Bleaches are used to decolourize materials, e.g. fabrics and paper. Sunlight and oxygen act as bleaches but the most commonly found bleach is sodium chlorate(I) solution (NaClO). This is produced when **chlorine** is reacted with **sodium hydroxide** solution:

$$Cl_2(g) + 2NaOH(aq) \rightarrow$$
$$NaCl(aq) + NaClO(aq) + H_2O(l)$$

Domestic bleaches such as 'Domestos' contain sodium chlorate(I) solution. The chlorate(I) decomposes to give oxygen which acts as an oxidizing agent:

NaClO + coloured $\rightarrow$ NaCl + oxidized
        material   (decolourized)
                                   material

Bleaching powder This is a white powder of complex composition. It is made by the action of **chlorine** on calcium hydroxide. It contains calcium chlorate(I) $Ca(OCl)_2$.

The powder releases chlorine when it is treated with dilute acid.

Boiling Boiling is the term used to describe the change from the **liquid** state to the **gas** state at a fixed **temperature** – the **boiling point**. Boiling takes place when the vapour pressure of the liquid equals the pressure of the atmosphere above the liquid.

Boiling point (temperature) This is the temperature at which **boiling** occurs. It is not fixed. It depends on the atmospheric **pressure**. The higher the pressure, the higher the boiling point. Boiling temperatures are also affected by impurities in the liquid. The presence of impurities causes the boiling temperature to rise. See **Evaporation**.

Bond Atoms are held together in **molecules** and **giant structures** by chemical bonds. These bonds are the forces which exist between the atoms and are generated by **electrons**. Electrons are either *shared between* atoms, or atoms *gain* or *lose* electrons forming ions.

The sharing of electrons (**covalent bonding**) occurs when similar atoms bond together. The loss or gain of electrons (**ionic** or electrovalent bonding) occurs when different types of atoms bond together, e.g. **metals** and **non-metals**.

the sharing of electrons
e.g. hydrogen (H_2)

the loss and gain of
electrons e.g. sodium
chloride ($NaCl$)

Boron Boron is a **non-metallic element** in **group III** of the **periodic table**. Its oxide is used to make **glass** – see **borosilicate glass**. It is a component of **alloys** and is a moderator in nuclear reactors.

Borosilicate glass The addition of boron oxide (B_2O_3) during the manufacture of **glass** results in a material which does not expand very much. Consequently the glass can be heated or cooled rapidly without cracking. 'Pyrex' glassware is like this.

Boyle's law At constant temperature the **volume** of a fixed **mass** of gas is inversely proportional to the **pressure** of the gas, i.e. if the pressure is *doubled* the volume is *halved*.

$$P \propto \frac{1}{V}$$

$$P.V = \text{constant}$$

pressure
increases
→

1P

volume I
=30 units

2P

volume II
=15 units

3P

volume III
=10 units

Brass Brass is an **alloy** of copper and zinc which usually contains about 30% zinc. Brass is an attractive yellow-golden coloured material which is used for ornaments and for electrical components, e.g. in electric plugs.

Brine A solution of sodium chloride in water. The solution is more concentrated than sea water and is used in the food processing industry, e.g. the production of bacon and in the preservation of many foods. In the chemical industry **chlorine** is produced by the **electrolysis** of **brine**.

Bromide Bromides are **compounds** of **bromine** and another **element**.

Metal bromides are usually **ionic** solids, e.g. sodium bromide NaBr. See **Halides**.

Non-metal bromides are usually **covalent** compounds, e.g. hydrogen bromide HBr.

33

Bromine (Br_2) Bromine is a member of the **halogen** group of elements. It is a **volatile** red liquid at room temperature. The liquid is very corrosive and the vapour is both irritating and poisonous.

Although it is less reactive than **chlorine**, bromine will react vigorously with metals forming **bromides**:

$$Mg(s) + Br_2(l) \rightarrow MgBr_2(s)$$

Bromine is extracted from sea-water by treatment with **chlorine**. One thousand litres of sea-water contains about 75g of bromine. Its chief uses are in the production of silver bromide for use in photo-

$C_2H_4Br_2$

graphic films and in the manufacture of 1, 2–dibromoethane which is added to **petrol** to prevent engine cylinders being blocked by deposits of lead. See **Octane rating**.

Bronze The combination of **copper** (>90%) with **tin** (<10%) results in an **alloy** which is much stronger than copper. The discovery of bronze in the Middle East (3000 B.C.) gave rise to important changes in the way man lived (The Bronze Age). Nowadays, bronze is mainly used for gear wheels and engine bearings and, of course, as medals e.g. for the Olympic Games.

Brownian motion Particles suspended in a **liquid** or **gas**, e.g. pollen in water or smoke in air, are seen to move in a random way. The explanation is that the microscopic particles which make up the liquid or gas, e.g. water and air **molecules**, are moving randomly and are hitting against the larger particles of pollen or smoke, causing them to move. Brownian motion is put forward as evidence for the **kinetic theory**.

Brown ring test This is a test for **nitrates**. If **concentrated sulphuric acid** is carefully poured

aqueous layer
brown layer
concentrated acid
(dense liquid)

into an **aqueous solution** of a nitrate to which an acidic solution of iron(II) sulphate has been added, a brown ring forms at the junction of the two layers of liquid. The nitric acid which is formed reacts with the iron(II) sulphate and produces a brown compound.

Burette A burette is a long glass tube with a tap on the end. The tube is marked (graduated) every 1 cm^3 with smaller graduations of 0.1 cm^3 as well. It is used in **titrations** to measure accurately small volumes (up to 50 cm^3) of liquid. The volume under the last graduation is called the dead space because it is not used for measurements. See **Pipette**.

Burner This device (sometimes called a bunsen burner) was invented in the nineteenth century so that coal-gas could be burnt cleanly giving a hot flame. **Natural gas** is now used. With the air hole open the gas is mixed with air and this makes sure that all the gas is burnt and no soot is produced. With the air hole shut a yellow sooty flame is produced because the gas is not fully burnt. See **Flame**.

flame

collar — jet
air
hole
← gas flows
in

Burning   See **Combustion**.

Butane   Butane is an **alkane** compound. It is a

$C_4 H_{10}$

gas at room temperature but it is easily liquefied. It is present (dissolved) in **petroleum** and is used in refineries both as a fuel and as a starting material for the production of **hydrogen** and **petrol** components. It is a major component of 'Calor gas'. The gas burns in air:

$$2C_4H_{10}(g) + 13O_2(g) \rightarrow 8CO_2(g) + 10H_2O(g)$$

By-product A by-product of a reaction is something which is produced in addition to the product which is required. For example, slag is a by-product of making **iron** in the **blast furnace** and **carbon dioxide** is a by-product of the fermentation process. It is important, if processes are to be economic, that by-products are sold, e.g. slag is sold as road making material.

Calcium Calcium is a soft, metallic **element** which is in **group** II of the **periodic table**. It is a fairly reactive **metal**, giving a slow but steady stream of **hydrogen** when it is added to cold water. 3.6% of the earth is made up of calcium compounds.

38

Many of these are very common, e.g. **limestone** and **chalk** ($CaCO_3$). Calcium is obtained by the electrolysis of molten calcium chloride.

Calcium compounds

Calcium carbonate $CaCO_3$	This is used as building stone and in the production of **cement**. **Lime** (calcium oxide) is produced from calcium carbonate in a *lime kiln*. See **Hardness of water**.
Calcium chloride $CaCl_2$	Used to produce calcium metal. The **anhydrous salt** is used as a drying agent.
Calcium hydrogen-carbonate $Ca(HCO_3)_2$	This is an important cause of temporary **hardness** in water.
Calcium hydroxide (slaked lime) $Ca(OH)_2$	This is a slightly soluble alkali. The aqueous solution is called **limewater**. It is used to make mortar.
Calcium oxide CaO	This is **lime**. It has very important uses in agriculture to combat excess acidity.
Calcium sulphate $CaSO_4$	Calcium sulphate gives rise to permanent hardness in water. It is found as anhydrite and gypsum.

Calor gas This is a mixture of **hydrocarbon** compounds which is used as a portable supply of gas. It is made up of mainly **propane** (C_3H_8) and **butane** (C_4H_{10}) and these compounds are stored in metallic bottles under **pressure**.

Carbohydrate Carbohydrates are **organic** compounds which contain the **elements, carbon, hydrogen** and **oxygen** only. Their formulae are always of the form: $C_x(H_2O)_y$, e.g. **sugars** such as **glucose** ($C_6H_{12}O_6$) **sucrose** ($C_{12}H_{22}O_{11}$) and ribose ($C_5H_{10}O_5$) and **polymers** such as **starch** ($C_6H_{10}O_5)_n$ and **cellulose** ($C_6H_{10}O_5)_n$. See **Monosaccharide, Disaccharide** and **Polysaccharide**.

Carbon Carbon is a **non-metallic element** which is in **group**(IV) of the **periodic table**. It is found in nature as two **allotropes**:
diamond★ and **graphite**.

All living tissue contains carbon compounds (**organic** compounds), e.g. **carbohydrates, proteins**, fats, and without these compounds life would be impossible.

Although the element burns in **oxygen** or **air** it is otherwise fairly unreactive. It is an important reducing agent. **Coke, coal** and **charcoal** are all impure forms of carbon. Carbon is also found combined in materials such as **limestone**.

Carbon compounds

Carbon dioxide CO_2

This gas can be formed by the action of heat or **acids** on **carbonates** or by the complete combustion of carbon. It is also produced in the **fermentation** of **sugars**. The test for carbon dioxide is that it turns **limewater** milky.
Carbon dioxide plays a vital role in **photosynthesis** and the **carbon cycle**.

Carbon monoxide CO

This gas is formed when carbon or its compounds are not completely burnt. It is an **air** pollutant, being produced in internal combustion engines and from the burning of cigarettes. It is a very poisonous gas because it combines with the **haemoglobin** in blood. It is useful as a **reducing agent** in the **blast furnace**. The gas burns to give **carbon dioxide**.

Carbon cycle   Everytime an animal breathes or a fuel is burnt carbon dioxide is released into the **atmosphere**. The process of **photosynthesis** continually takes carbon dioxide out of the atmosphere and builds up plant structures with it. In these processes carbon dioxide is cycled round the earth.

There are stores of it (dissolved in the oceans, lakes and rivers) in **carbonate** rocks and in **fossil fuels**. A simple diagram of these movements is shown here. Some changes occur quickly, e.g. burning, others very slowly, e.g. formation of rocks.

Carbon dating Carbon has a **radioactive isotope** ($^{14}_{6}C$) and the atmospheric **carbon dioxide** contains a small constant proportion of this isotope. All carbon compounds in living tissue contain this carbon-14 isotope in the same constant proportion. When an animal or plant dies, however, the proportion falls at a known rate. It falls because plants no longer take in the isotope through **photosynthesis** and animals no longer take in the isotope by eating plants. All the time the isotope is decaying. By measuring the amount of radioactivity in a dead material it is possible to estimate the age of the material. The technique is used by archaeologists.

Carbonate ($-CO_3$) The carbonate group has a **valency** of two. Metal carbonates occur widely in nature, e.g. **limestone** ($CaCO_3$), dolomite ($CaMg(CO_3)_2$), and malachite ($CuCO_3.Cu(OH)_2$). Calcium carbonate plays a part in the **carbon cycle**.

Carbonates of all **metals** (except **group** I metals) are **insoluble** in water. All carbonates produce **carbon dioxide** when heated *strongly* (K_2CO_3, Na_2CO_3 with difficulty) and when treated with dilute acid. The chemical test of a carbonate is to add acid to the solid compound and to pass the gas produced through **limewater**. A **precipitate** indicates that the solid was a carbonate.

Catalysis See **Catalyst**.

Catalyst A catalyst is a substance which alters the **rate** of a chemical reaction. The catalyst remains unchanged at the end of the reaction. The process is called **catalysis**.

Transition metals and their compounds are often useful catalysts. Some examples are shown here:

43

Reaction	Catalyst
Haber process	Iron
Contact process	Vanadium(v) oxide (V_2O_5)
Ammonia → nitric acid	Platinum-rhodium alloy
Hardening fats	Nickel

The chemical processes which go on inside animals and plants are nearly all dependent on catalysts. These organic catalysts are called **enzymes**.

Cathode The negatively charged pole in a **battery** or **electrolysis cell** is called a cathode. Positively charged **ions (cations)** are attracted to the cathode during electrolysis. They gain electrons at the cathode.

Cation Cations are positively charged **ions**. Most cations are metal ions, e.g. Fe^{2+}, Cr^{3+}, Na^+, Ca^{2+} Al^{3+} and are attracted to the **cathode** during **electrolysis**. Three cations not formed from metals are the **hydrogen ion** (H^+) the **oxonium ion** (H_3O^+) and the **ammonium ion** (NH_4^+).

Caustic This term is used to describe substances which burn or corrode organic material, e.g. flesh. It is usually restricted to use with **alkaline** materials, e.g. caustic soda (**sodium hydroxide**), and caustic potash (**potassium hydroxide**).

Celcius scale of temperature (°C)

0°   37°   100°

melting point   usual   boiling point
of pure ice   temperature   of pure water
               of blood

This scale of temperature is based on a 100° range between the melting point of pure **ice** (0°C) and the boiling point of pure water (100°C). Originally called the *centigrade* scale, it is now named after Celcius, the man who devised it. One degree Celcius equals one **Kelvin**.

Cell 1) This term is used as an alternative name for a **battery**.

2) It is also used to describe the single units from

negative poles

electrolyte

positive poles

45

which large batteries are made, e.g. the lead-acid **accumulator** used in most cars contains six cells, each of which gives two volts. The total voltage is twelve volts as the cells are connected together in series in the accumulator. The common dry battery is a single cell.

3) The term is also used to describe apparatus and chemicals which produce an electrolysis, e.g. *electrolytic cell*.

Cellulose This is a **carbohydrate** which is a **polymer**. It is made up of **glucose monomers**. The formula is $(C_6H_{10}O_5)_n$ and is the material from which cell walls of plants are made. Cellulose is useful to man in two ways. It is used to manufacture paper and **rayon**.

Cement Cement is made by heating **limestone** and clay together and then powdering the product. When water is added and the mixture is allowed to dry a hard substance is produced.

cement + sand + water → mortar
cement + sand + gravel + water → concrete

Centrifuge If it is wished to separate a **solid** from a **liquid**, a centrifuge can be used. The mixture is placed into tubes and put into the machine. The tubes are spun round at very fast speeds and the solid material is forced to the bottom of each tube. Here it

⁻ forms a compact mass and the liquid can be poured off. Centrifuges are dangerous machines and the lid must always be closed when they are in use.

Chain Carbon atoms are able to **bond** together in chains and can form *long* **molecules**. In **polymers**, these chains can be thousands of atoms long. The chains can have branches too. An example of a chain is shown here:

C_5H_{12} – pentane

Chain reaction Some **isotopes** are unstable and when they are bombarded with a **neutron** they break up to produce smaller atoms and more neutrons (nuclear **fission**).

$$^{235}_{92}U + \text{neutron} \rightarrow \left\{\begin{array}{l}\text{atoms of } \textbf{barium}\\ \text{and } \textbf{krypton}\end{array}\right\} + 3\,\text{neutrons}$$

Each time a neutron is used up, three more are produced, each of which can then go on to split up another large unstable atom and produce even more neutrons. These can then go on and on and more and more neutrons are produced. This is called a chain reaction and is the basis for **nuclear reactions** in nuclear power stations and atom bombs.

Chalk This is rock which has been formed from the shells of marine animals. It is mainly **calcium carbonate** ($CaCO_3$). It is a softer rock than **limestone** and its uses are limited by this.

Charcoal Charcoal is made by heating **organic** material (usually wood) in the absence of **air**. In this process, volatile material escapes and the resulting charcoal is mainly composed of **carbon**. Before coal was widely available, charcoal was used to make iron from iron ore. Some charcoal which is called *activated charcoal* is able to absorb small molecules onto its surface. This is used in gas masks, in the **sugar** refining industry and has other uses where coloured impurities need to be removed from materials.

Charles' law The volume of a fixed mass of gas is dependent upon its temperature. If the temperature of a gas is doubled (measured on the **Kelvin** scale) then the volume of the gas will double – *if* the **pressure** of the gas is kept constant.

volume

8 cm³	16 cm³	24 cm³

200 K	400 K	600 K
(−73°C)	(127°C)	(327°C)

temperature

Chemical change This is a change in which one or more chemical substances are changed into *different* ones. Such a change is usually accompanied by the giving out or taking in of **heat energy**. See **Physical change**.

Chemical reaction See **Reaction**.

Chlorides These are **compounds** of chlorine and another element. Metal chlorides are usually **ionic** solids, e.g. sodium chloride ($NaCl$) and barium chloride ($BaCl_2$).

Non-metal chlorides are **covalent** compounds and are usually either low boiling point liquids such as tetrachloromethane (CCl_4) or gases such as hydrogen chloride (HCl).

Chlorides react with concentrated **acids** to produce **hydrogen chloride** gas:

$$H_2SO_4(l) + NaCl(s) \rightarrow NaHSO_4(s) + HCl(g)$$

The test for chlorides is to mix a solution with silver nitrate solution. If a white **precipitate** forms which then dissolves when **ammonia** solution is added the substance is a chloride.

Chlorination **Chlorine** is added to drinking **water** and to water used in swimming pools to kill dangerous bacteria. The term is also used to describe reactions between chlorine and **hydrocarbons** to produce chlorinated hydrocarbons.

Chlorine (Cl_2) Chlorine is a green **gas** at room temperature. It is a member of the **halogen** group of elements (**group** VII) and is very reactive. It has a choking effect and attacks lung tissue and the throat if it is breathed in. It was used as a chemical weapon in the 1914–18 war.

Chlorine occurs as **chlorides** and sodium chloride is abundant in sea water. Chlorine is extracted by the **electrolysis** of **rock salt** solutions. It is used widely to make **polymers (PVC)**, pesticides **(DDT)**, **disinfectants** (TCP) and solvents (dry cleaning). A solution of chlorine in water is used as a **bleach**.

Chlorine is a vigorous **oxidizing agent** which readily reacts with most **elements**. It is made in the laboratory by the oxidation of concentrated hydrochloric acid, e.g:

$$4HCl(aq) + MnO_2(s) \rightarrow$$
$$Cl_2(g) + MnCl_2(ag) + 2H_2O(l)$$

Chloroethene (vinyl chloride)

C_2H_3Cl

Chloroethene is the monomer from which **poly-(chloroethene)** (PVC) is made.

It is made by reacting ethene (C_2H_4) with chlorine (Cl_2).

Chlorophyll Chlorophyll is the green pigment in plants which is responsible for the absorption of the sun's energy by the plant. This energy is used in the process of **photosynthesis**. There are two chlor-

51

ophyll molecules each of which is a complex molecule which contains a **magnesium** atom.

Chromatography This is a technique for separating mixtures of **solutes** in a **solution**. One simple method of separating two pigments is to put a spot of the mixture onto the centre of a filter paper (I) and then to add drops of water to the spot. As the water moves out across the paper the pigments begin to separate (II). Eventually two quite separate bands of coloured material will be seen on the paper. The inks in fibre-tipped pens can be separated in this way. The process occurs because some materials cling to the paper more than others, e.g. in the example pigment B is more tightly held than A.

Chromium Chromium is a **transition metal**. It has important uses in **alloys**, e.g. **stainless steel** and is used in chromium plating on items such as bicycle handlebars, kettles, car bumpers and cutlery. It has a high resistance to corrosion. See **Dichromate**(VI).

Citric acid   This is a **weak organic acid** which occurs naturally in *citrus* fruits such as oranges and lemons.

cm³   This is the abbreviation for cubic centimetre and is a unit of **volume** used in scientific work. It is the volume of a cube which has a side of 1 cm. One thousand cm³ are equal to one litre or one cubic decimetre (1 **dm³**).

Coal   Coal is fossilized plant material which was living millions of years ago. It has a very complex chemical structure containing compounds made up of **carbon, hydrogen, oxygen, nitrogen** and **sulphur**. Coal is used as a fuel in power stations, industry and the home and was, before the use of **natural gas**, used as the source of gas (coal gas). About 20% of coal is used to make **coke**.

If coal is heated in the absence of air, coal-tar is produced. In the middle of this century coal-tar was an important source of **organic** chemicals, e.g.

phenol from which dyes, drugs and **polymers** were made. Now over 90% of such products come from **petroleum** sources.

As the world has greater reserves of coal than petroleum, attempts are being made to convert coal into petroleum products.

Cobalt Cobalt is a **transition element**. It is a magnetic element and is used alloyed with iron in the manufacture of magnets. The radioactive **isotope** $^{60}_{27}$Co emits **gamma rays** and it is used in the treatment of cancers. Cobalt(II) chloride is blue when **anhydrous** and pink when **hydrated**, and so the compound is used as a test for the presence of water.

Coke Coke is the material left behind when the **volatile** compounds have been removed from **coal**. It contains over 80% carbon. It was made during the production of coal-gas but now is made in *coke ovens* from special *coking coal*. Coke is used to make **carbon monoxide** in the **blast furnace** and is an important (smokeless) fuel.

Combining power This is another term for **valency**.

Combustion In combustion (or burning) reactions, a substance combines with a **gas**. Heat and light (i.e. **flame**) usually accompany combustion reactions. Most combustions involve **oxygen**, e.g:

$$2H_2(g) + O_2(g) \rightarrow 2H_2O(g)$$

Complex ions Complex ions consist of a **cation** (e.g. metal ions or hydrogen ions) which is bonded to one or more small molecules by a **co-ordinate bond**.

The simplest examples are the ammonium ion: and the oxonium ion:

$$^+H \leftarrow :N\!-\!H$$

with H above and H below

$$NH_4^+$$

$$H^+ \leftarrow :O$$

with H above and H below

$$H_3O^+$$

Here the **lone pair of electrons** on the nitrogen and oxygen atoms forms the bond.

Transition metal ions form a large number of complex ions because they can easily accept the donated electron pairs.

Compound A compound is a pure substance which is made up of two or more elements chemically bonded together. The **properties** of a compound are totally different from the properties of the elements bonded together within it. Compounds may contain **ionic** or **covalent** bonding. Examples: methane CH_4, water H_2O, sodium chloride NaCl.

See **Mixture**.

Concentrated A concentrated solution is one which contains a relatively large concentration of solute. If you want to increase the concentration of a solution, this can be done by either adding more solute or by removing solvent (e.g. by **distillation**) from the solution. See **Dilute**.

Concentration The concentration of a **solution** is a measure of how much **solute** is **dissolved** in the solution. It is usually expressed in terms of how much substance is present in a given volume of the solution. This can either be in terms of **mass**, e.g. grams, or in terms of how many particles, e.g. **moles**. Examples of concentrations and how they can be written are:

170 grams of silver nitrate (1 mole) in one cubic decimetre (dm^3) of solution:

$$170 \text{ g/dm}^3, \text{ or } 1 \text{ mol/dm}^3$$

Condensation 1) The change from the **gas** or **vapour** state to the **liquid** state, e.g. water condensing onto a cold window pane.

2) A reaction in which two or more molecules react to produce a larger molecule and a small molecule such as water. This is one method of producing **polymers**, e.g. **nylon**.

Conductor An *electrical* conductor is a substance which will allow an electric **current** to flow through it. **Metals**, solutions which contain **ions** and molten **ionic** compounds are conductors; all other substances are **insulators**. Certain **metalloids** are **semiconductors**.

Examples of conductors are:

Sodium (Na). Conduction by **electrons**.	Molten copper(II) chloride (CuCl$_2$(l)). Zinc chloride solution (ZnCl$_2$(aq)). Conduction by movement of **ions**.

The term conductor is also applied to substances which allow **heat energy** to flow through them. Metals are also good conductors of heat.

Conservation of energy and mass In a chemical reaction the **mass** of the reacting substances and the mass of the products are equal. Similarly:

Energy possessed by reacting substances.	=	Energy possessed by the products of the reaction.	+	Any **heat energy** which is exchanged with the surroundings.

In other words mass and energy cannot be created or destroyed in a chemical reaction. They are *conserved*.

These relationships are only approximations. The work of Einstein showed that mass and energy can be converted into each other. This is important in **nuclear reactions** but in ordinary chemical reactions the relationship holds good.

Contact process This process is used to make almost all the **sulphuric acid** produced today. The flow chart shows the steps:

sulphur ⌐ $S + O_2 \rightarrow SO_2\star$

or

sulphide ores (e.g. ZnS) — reaction with **air** → sulphur dioxide SO_2 — reacted with air over a vanadium (v) oxide **catalyst** to produce:

$2ZnS + 3O_2 \rightarrow 2ZnO + 2SO_2$

sulphuric acid H_2SO_4 (98%) ← sulphur trioxide gas is absorbed$\star$ into concentrated sulphuric acid (98%) which is then diluted with water to keep the concentration at 98% — sulphur trioxide SO_3 ← 450°C

$2SO_2 + O_2 \rightarrow 2SO_3\star$

Most sulphuric acid is now made from **sulphur**. The reactions marked ★ are **exothermic** and so the process is relatively cheap to run because little heat energy has to be supplied.

Co-ordinate bond This is a **covalent bond** in which *both* the **electrons** in the bond have come from the *same* **atom**. Electrons have been *donated* by one atom rather than one donated from each atom. The bond is also known as a **dative bond**. Examples are:

In co-ordinate bonding, the donor atom uses a **lone pair of electrons**. See **Complex ion**.

Copper Copper is a **transition element** which plays an important part in our lives. It is a vital trace element in our bodies – we need between one and two milligrammes per day.

Copper is an unreactive metal, coming low down in the **electrochemical series**. It will not liberate **hydrogen** from dilute **acids**.

It can be purified by using the impure metal as the **anode** with copper(II) sulphate as the **electrolyte** in an **electrolysis cell**.

Copper metal is used in plumbing and for electrical wiring. (Copper is an excellent **conductor**.) It is extracted commercially from its **sulphide** ores:

Chalcopyrite   ($CuFeS_2$)
Bornite   (Cu_5FeS_4)
and its carbonate ores:
Malachite   ($CuCO_3.Cu(OH)_2$)
Azurite   ($2CuCO_3.Cu(OH)_2$)

Copper compounds

Copper(II) chloride $CuCl_2.2H_2O$	Copper(II) chloride is used to give fireworks a green colour and in the removal of **sulphur** from **petroleum**.
Copper(II) hydroxide $Cu(OH)_2$	Copper(II) hydroxide is used in the manufacture of **rayon**, in dyeing textiles and as a blue pigment.
Copper(II) carbonate $CuCO_3$	Copper(II) carbonate forms part of the ores malachite and azurite – both basic copper carbonates. Another basic carbonate –

verdigris is formed as a surface layer on copper which is exposed to the **atmosphere**.

| Copper(ii) sulphate $CuSO_4.5N_2O$ | Copper sulphate forms blue crystals and can be made by treating copper compounds (oxide, carbonate, hydroxide) with dilute sulphuric acid. The **anhydrous** salt is a white powder which turns blue when water is added: |

$$CuSO_4 + 5H_2O \rightarrow CuSO_4.5H_2O$$
white   blue

This reaction is often used as a test for the presence of water. The compound is used in wood preservatives and fungicides, e.g. Bordeaux mixture.

Corrosion When a substance corrodes it is chemically eaten away. Good examples are the weathering of **limestone** buildings by rainwaters which contain dissolved acids; **rusting** of **steel**. Corrosion begins at the surface and often a surface layer protects the rest of the material, e.g. the **oxide** coating on **aluminium**. Rusting is not like this, the rusting process goes through the steel until it is all corroded. That is why it is so damaging and expensive.

Coulomb *Electric charge* is measured in coulombs. It is the product of the **current (units = ampère)**

and the time (units = **seconds**), e.g. if a current of 0.6 amps flows for ten seconds, six coulombs of electricity will have passed through the circuit.

Covalent bonds These **bonds** are formed by two **atoms** coming together and sharing their **electrons**, e.g. **hydrogen** atoms have one electron each (●). The atoms form a **molecule** of hydrogen (H_2) by the two electrons forming a bond:

hydrogen atoms (H)

hydrogen molecule (H_2)

hydrogen chloride

Covalent bonds are usually found in compounds which only contain **non-metallic elements**, e.g.

Carbon dioxide	CO_2
Hydrogen chloride	HCl
Nitrogen dioxide	NO_2
Ammonia	NH_3
Phosphorus(v) oxide	P_2O_5

and also in elements with an **atomicity** of two or more, e.g.

Oxygen	O_2	Hydrogen	H_2
Bromine	Br_2	Chlorine	Cl_2
Nitrogen	N_2	Sulphur	S_8
Ozone	O_3	Phosphorus	P_4

Covalent bonds which contain two electrons are termed *single* bonds. Many molecules contain **double** or **triple** bonds. Covalent bonds are not as strong as **ionic bonds**.

Covalent compounds These are compounds which contain **covalent bonds**. They tend to be non-**conductors**. Their **melting** and **boiling points** are low.

Cracking This is the process in which *large* **hydrocarbon molecules** are broken up into *small* molecules. This occurs in the **petroleum** refining industry. Small molecules are more valuable than large ones because they are the starting materials needed in the production of such things as **polymers**, e.g. **poly(ethene)** and **poly(propene)**, and fuels, e.g. **petrol** and **diesel fuel**.

Different products are formed under different conditions:

Conditions	Process name	Products
High temperature.	Thermal cracking.	**Unsaturated** and **saturated** molecules are produced.
High temperature in the presence of steam.	Steam cracking.	
High temperature in the presence of a catalyst.	Catalytic cracking.	
High temperature and pressure in the presence of hydrogen.	Hydro-cracking.	The product is totally saturated.

Cryolite In the extraction of **aluminium, bauxite** (Al_2O_3) is dissolved in molten cryolite (Na_3AlF_6) at 900°C and the mixture is then electrolysed. Aluminium is produced at the **cathode**.

Crystal **Solid** substances have their atoms arranged in a set geometrical pattern. The shape of these atomic arrangements (crystal **lattices**) deter-

cubic

octahedral

tetrahedral

mines the shapes that the solid has when it undergoes **crystallization**. Some examples of different crystal arrangements:

Crystallization This is the process in which **crystals** are formed. In nature, crystals are produced when molten rocks cool down and solidify. In laboratory experiments, crystals are usually produced from a **solution**. There are two main methods:

1) A solution is left at **room temperature**. Slow **evaporation** of the **solvent** takes place and crystals are left behind.
2) A **supersaturated solution** is made above room temperature. As this is cooled down, crystals are produced.

Current Electric current is the movement of **electrons** through a **conductor**. It is measured in **ampères** (amps). Current is said to move from positive terminal to negative terminal. The electrons, however, travel from the negative terminal to the positive. This difference is because the electron carries a negative electric charge.

Cyanide Compounds which contain the $-CN$ group of atoms are known as cyanides. Potassium cyanide (K^+CN^-) and hydrogen cyanide ($H - C \equiv N$) are the best known examples. They are very poisonous. Cyanides prevent the flow of **energy** within the body and so are effective very quickly.

Decant To pour off. If a **precipitate** is allowed to settle, the **liquid** remaining can then be decanted by pouring it carefully away.

Decomposition The break up of compounds into simpler compounds or into elements. Usually heat is needed (thermal decomposition), e.g:

$$2Cu(NO_3)_2(s) \rightarrow 2CuO(s) + 4NO_2(g) + O_2(g)$$
$$2HgO(s) \rightarrow 2Hg(l) + O_2(g);$$
$$CaCO_3(s) \rightarrow CaO(s) + CO_2(g)$$

Decrepitation When some **crystals** are heated a sharp crackling noise is heard. This is the sound of the crystals breaking up. Lead(II) nitrate ($Pb(NO_3)_2$) is an example of a salt which undergoes decrepitation.

Dehydrating agent A dehydrating agent is a substance which is used to remove water from other substances. Such substances always have an attraction for water. They can be substances which dissolve in water, e.g:

Concentrated sulphuric acid H_2SO_4	Sodium hydroxide NaOH

or substances which react with water, e.g:

Calcium oxide CaO	this forms the hydroxide $\rightarrow Ca(OH)_2$

A third kind of dehydrating agent is the **anhydrous** salt which absorbs water, e.g:

Calcium chloride $CaCl_2$ $\rightarrow CaCl_2.2H_2O$	Sodium sulphate Na_2SO_4 $\rightarrow Na_2SO_4.10H_2O$

Dehydrating agents are used in **desiccators**.

DDT

Dichloro**d**iphenyl**T**richloroethane.

DDT is a widely used pesticide which has been successful in the control of diseases such as *malaria*. It is, however, harmful to animal life as some animals can concentrate it in their bodies where it acts as a **poison**. For this reason some countries restrict its use.

Deliquescence Deliquescent substances are able to take in water from the **atmosphere**. They can take in so much water that they are able to form a solution – unlike **hydroscopic** substances.

Some common examples are:

Iron(III) chloride $FeCl_3$	Copper(II) nitrate $Cu(NO_3)_2$	Calcium chloride $CaCl_2$

Some deliquescent substances are used in **desiccators**.

Delocalized electrons In normal **covalent bonds, electrons** are found between two atoms, e.g. a carbon–carbon **single bond** consists of two electrons:

$$C : C$$

and a double bond contains four electrons:

$$C :: C$$

These electrons are located in between the atoms. They are *localized*. In some substances, however, there are some electrons which are free to move amongst the atoms. These electrons are *de*localized. **Graphite** contains delocalized electrons *within* the hexagonal layers of atoms. This is why graphite can conduct electricity along the layers *but not* between them. **Metals** have delocalized electrons too and this explains their being good **conductors**.

Density The density of a substance is a measure of how much space a certain amount of the substance takes up. The smaller the space into which the **mass** is concentrated, the *greater* is the density. The density of a substance, e.g. lead, is determined by the density of the individual **atoms** *and* by how closely together the atoms are arranged in the **crystal lattice**. Generally speaking, the more particles that an atom contains, the denser it will be.

Density is calculated by dividing a substances mass by its volume. The units are usually either g/cm^3 or kg/m^3.

Desiccator Desiccators are used to store substances which need a dry atmosphere, e.g. **deliquescent** or **hygroscopic** substances.

A desiccator is air-tight and contains a **dehydrating agent** which makes the air dry.

Detergent A detergent is a cleaning agent. It does two things when used in washing, e.g. clothes or dishes. It reduces the surface tension of the water and so allows the water to wet things more thoroughly. It also acts to bring together the water and (normally insoluble) fat, **oil** or grease into an emulsion. If the presence of a detergent is accompanied by rapid movement, e.g. in a washing machine, then particles of dirt and grit are removed and the article is cleaned.

Detergents form an emulsion because one part of their molecule is **ionic** and is attracted to water and because the other part (the **hydrocarbon** chain) is attracted to the oil molecules. Detergents, thus, hold the two together. An example of a synthetic detergent molecule:

$$H-\underset{\underset{H}{|}}{\overset{\overset{H}{|}}{C}}-\underset{\underset{H}{|}}{\overset{\overset{H}{|}}{C}}-\underset{\underset{H}{|}}{\overset{\overset{H}{|}}{C}}-\underset{\underset{H}{|}}{\overset{\overset{H}{|}}{C}}-\underset{\underset{H}{|}}{\overset{\overset{H}{|}}{C}}-\underset{\underset{H}{|}}{\overset{\overset{H}{|}}{C}}-\underset{\underset{H}{|}}{\overset{\overset{H}{|}}{C}}-\underset{\underset{H}{|}}{\overset{\overset{H}{|}}{C}}-\underset{\underset{H}{|}}{\overset{\overset{H}{|}}{C}}-\underset{\underset{H}{|}}{\overset{\overset{H}{|}}{C}}-\underset{\underset{H}{|}}{\overset{\overset{H}{|}}{C}}-\underset{\underset{H}{|}}{\overset{\overset{H}{|}}{C}}-\underset{\underset{O}{\|}}{\overset{\overset{O}{\|}}{S}}-\overset{-}{O}\,\overset{+}{Na}$$

See **Soap, Hydrophilic** and **Hydrophobic**.

Deuterium This is an **isotope** of hydrogen which has a **neutron** and a **proton** in the **nucleus**. Its symbol is written

$$^2_1H \text{ or } ^2_1D.$$

The isotope occurs naturally, making up 0.015% of hydrogen. Because its density is twice that of normal hydrogen it is easily separated.

Deuterium oxide is known as *heavy water* and its formula is written D_2O.

Di- This is a prefix which means *two*, e.g:

Carbon *di*oxide CO_2 – two atoms of oxygen.
*Di*atomic molecule – two atoms in the molecule.

Diamond This is an **allotrope** of **carbon**. It occurs naturally – mainly in South Africa and the Soviet Union and is a valuable mineral. It is the hardest naturally-occurring substance known and is used in drill tips and saw blades. Diamond is prized as a gemstone because of its rarity and the sparkle it produces. The carbon **atoms** are arranged in the **crystal lattice** in a **tetrahedral** formation and are covalently bonded to each other.

Diatomic molecule A **molecule** which contains two atoms is said to be **di**atomic, e.g.

N_2 $\qquad$ O_2 $\qquad$ H_2 $\qquad$ Cl_2 $\qquad$ F_2 $\qquad$ Br_2 $\qquad$ I_2.

Dibasic acid An **acid** which contains two **hydrogen** atoms per molecule is called a **di**basic acid.

Some examples are shown. With such acids, two kinds of salt can be formed – the *normal* salt, in which *both* hydrogen atoms are replaced, and the **acid salt** in which only one hydrogen has been replaced. Examples of these are also shown.

Acid	Salts	
	acid	*normal*
H_2SO_4 sulphuric acid	$NaHSO_4$ sodium hydrogen-sulphate	Na_2SO_4 sodium sulphate
H_2CO_3 carbonic acid	$NaHCO_3$ sodium hydrogen-carbonate	Na_2CO_3 sodium carbonate

Dichromate(VI) ion $(Cr_2O_7^{2-})$ This ion contains **chromium** and **oxygen** atoms and has a valency of two. Its formula is $Cr_2O_7^{2-}$ and it is usually used as the bright orange potassium or ammonium salts:

$$K_2Cr_2O_7 \quad (NH_4)_2Cr_2O_7$$

Dichromate ion is an **oxidizing agent** which is reduced to chromium(III) ions Cr^{3+}.

Diesel fuel A fuel/air mixture is injected into diesel engines where it is compressed. The temperature produced in the compression causes the fuel/air mixture to explode. The fuel is produced from **petroleum** and is made up mainly of alkanes in the boiling range 200°C–350°C. It is often sold as 'DERV' fuel:

Diesel
Engine
Road
Vehicle

Diffusion If two gases (e.g. air and nitrogen dioxide) are at first separated and then allowed to mix, they mix together *completely*. This happens

before   after

because the molecules in the gases are moving about because of the thermal **energy** they possess.

Diffusion occurs at a faster rate when the temperature is raised. The *less* dense a gas is, the *greater* is its rate of diffusion.

Diffusion also occurs in liquids, solutions and solids.

Dilute A dilute **solution** is one which contains a relatively low **concentration** of **solute**. If you want to make a solution *more* dilute it is usual to add more solvent to it. Dilute acids usually have concentrations of 2 **mol/dm³** *or less*.

Dimer A dimer is a molecule which is made up of two identical molecules (monomers) which are bonded together.

Example:

N_2O_4 is a dimer of NO_2

Disaccharide A disaccharide is a **sugar** molecule which is made up of two **monosaccharide** sugar molecules which have undergone a **condensation** reaction with the elimination of a molecule of water.

Examples:

Monosaccharides		Disaccharide
glucose + glucose		maltose
$C_6H_{12}O_6$ $C_6H_{12}O_6$		$C_{12}H_{22}O_{11}$
glucose + fructose		sucrose
$C_6H_{12}O_6$ $C_6H_{12}O_6$		$C_{12}H_{22}O_{11}$

Disinfectant A disinfectant is a substance which is capable of destroying harmful bacteria. They are often based on the compound **phenol**; two common household examples are shown here:

used in
T.C.P.

used in
Dettol

Displacement reactions In a displacement reaction a less reactive element is displaced by a more reactive one. These are **redox** reactions.

Example:

$$CuSO_4(aq) + Zn(s) \rightarrow ZnSO_4(aq) + Cu(s)$$
$$Mg(s) + CuO(s) \rightarrow MgO(s) + Cu(s)$$
$$2NaBr(aq) + Cl_2(g) \rightarrow 2NaCl(aq) + Br_2(aq)$$

Dissociation When a compound dissociates it breaks up into smaller simpler molecules or ions. The dissociation is sometimes **reversible**.

$$NH_4Cl(s) \underset{cool}{\overset{heat}{\rightleftharpoons}} NH_3(g) + HCl(g)$$

ammonium   ammonia   hydrogen
chloride   chloride

$$HCl(aq) \rightleftharpoons H^+(aq) + Cl^-(aq)$$

Dissolve When a **solute** dissolves in a **solvent** to form a **solution** it is dispersed through the whole volume of the solvent. The structure of the solute is broken up.

Distillation This is the process by which a **solvent** can be recovered from a **solution** or a **mixture**. The solution is heated and the solvent is turned into a vapour. The vapour is led away from the heat and is then condensed and collected. In this way pure solvent can be recovered. A typical set of apparatus is shown.

See **Fractional distillation**.

Distilled water Tap water and rain water are not **pure**. They contain dissolved **salts** and **gases**. Water is often distilled to increase its purity. Most of the salts are left behind but the water still may contain dissolved gases. The presence of **carbon dioxide** reduces the **pH** of the water considerably.

dm³ This is the symbol for cubic decimetre, 1000 **cm³**.

DNA **D**eoxyribo**n**ucleic **a**cid is the very complex chemical which controls the order in which **amino acids** combine to produce **proteins**.

Double bond A chemical **bond** which contains *two shared pairs* of **electrons** is a double bond. Carbon–carbon double bonds are found in **alkene** molecules. The presence of the extra electrons makes the alkenes very reactive. Such compounds which contain double (and **triple**) bonds are said to be **unsaturated** compounds.

Compounds and elements containing double bonds:

Oxygen $\qquad$ O=O

Ethene

$$\begin{array}{c} H \\ \diagdown \\ C=C \\ \diagup \quad \diagdown \\ H \qquad H \end{array}$$

Carbon dioxide $\qquad$ O=C=O

Double decomposition This process occurs when ionic substances react in solution. The product is an insoluble solid.

Example:

$$CaCl_2(aq) + Na_2CO_3(aq) \rightarrow$$
$$CaCO_3(s) + 2NaCl(aq)$$
$$\text{or } Ca^{2+}(aq) + CO_3{}^{2-}(aq) \rightarrow CaCO_3(s)$$

This is a method of preparing insoluble **salts**.

Dry cell This is a **battery** in which the electrolyte is in the form of a paste and so cannot spill from the battery.

The only common *wet* battery used today is the lead-acid **accumulator**.

Dry ice This is the name given to **solid carbon dioxide**. This material **sublimes** and hence the name *dry*. It is used for keeping things cold (sublimation point = $-40\,°C$) and in the theatre for producing smoke and clouds on stage.

Ductile A ductile substance can easily be drawn into wires, e.g. **copper** is a very ductile **metal**. The term is usually used only about metals. Ductile metals have large **crystals** within them. See **Annealing** and **Malleable**.

Duralumin This is an **alloy** of **aluminium**. It contains **copper** (4%) and traces of **manganese, magnesium** and **silicon**. The alloy is much *harder* and *stronger* than the pure metal. It is used for the production of aircraft frames and cooking utensils.

Dyes These are chemicals which can be mixed or reacted with materials to make a coloured product, e.g. fabrics, plastics. The colour is produced because the dye absorbs some of the light which falls upon it and radiates the rest.

Dynamite This explosive was invented by Alfred Nobel. He discovered that if the unstable explosive nitroglycerine was absorbed into a clay called kieselguhr the result was a *stick* of explosive which was safe until it was detonated.

$$H - \overset{\overset{\displaystyle H}{|}}{\underset{\underset{\displaystyle NO_3}{|}}{C}} - \overset{\overset{\displaystyle H}{|}}{\underset{\underset{\displaystyle NO_3}{|}}{C}} - \overset{\overset{\displaystyle H}{|}}{\underset{\underset{\displaystyle NO_3}{|}}{C}} - H$$

Nitroglycerine

Efflorescence Efflorescence is the process in which crystals lose part of their **water of crystal-**

lization when they stand in the air. They go powdery on their surfaces. A well-known example of this process is sodium carbonate (washing soda):

$$Na_2CO_3.10H_2O(s) \rightarrow Na_2CO_3(s) + 10H_2O(g)$$

Electric current or **Electricity**   See **Current**.

Electrochemical series   **Elements** differ in their reactivity. When a metal is dipped into a **solution** of one of its salts, e.g. zinc in zinc sulphate solution, the following reaction occurs:

$$zinc \rightarrow zinc \ ions + 2 \ electons$$
$$Zn(s) \rightarrow Zn^{2+}(aq) + 2e^-$$

The ions go into solution and the **electrons** cling to the metal. If the metal and solution are made part of a circuit the electrons will flow round the circuit. The *more* reactive a metal is the greater the force with which the electrons move.

K	potassium	↑ reactivity *increases*
Na	sodium	
Ca	calcium	
Mg	magnesium	
Al	aluminium	
Zn	zinc	
Fe	iron	
Pb	lead	
H	hydrogen	
Cu	copper	
Ag	silver	

When metals are put in order of the force they produce it is known as the electrochemical series:

The series is useful because it allows us to predict reactions, e.g:

zinc *will* remove the oxygen from copper(II) oxide, but

copper *will not* remove oxygen from zinc oxide;

hydrogen *will* reduce copper(II) oxide *but not* zinc oxide

copper *will not* react with acids to release hydrogen.

Electrode An electrode is a **conductor** which dips into an **electrolyte** and allows the **current (electrons)** to flow to and from the electrolyte. **Copper, carbon** and **platinum** are used in **electrolysis** as electrodes. In electrolysis chemical reactions occur *at* the electrodes. See **Anode, Cathode**.

Electrolysis When a *direct* electric current is passed through a liquid which contains ions (an

Anode ⊕

Cathode ⊖

negative ions (**anions**) e.g. Cl^-

positive ions (**cations**) e.g. H^+

Electrolyte (sodium chloride solution)

electrolyte) chemical changes occur at the two **electrodes**.

at the anode	at the cathode
$Cl^- \rightarrow Cl + e^-$	$H^+ + e^- \rightarrow H$
$2Cl \rightarrow Cl_2$	$2H \rightarrow H_2$
chlorine is given off	hydrogen is given off

$$2[H^+ + Cl^-] \rightarrow H_2 + Cl_2$$

Electrolysis is used to extract metals and non-metals from compounds. Aluminium, sodium and copper are produced in this way, as is chlorine.

Electrolyte An electrolyte is either a molten ionic compound:

$$NaCl(l) \quad PbI_2(l)$$

or a solution which contains ions:

$$HCl(aq) \quad NaOH(aq) \quad KI(aq) \quad CuSO_4(aq)$$

Chemical changes take place when a direct electric **current** is passed through an electrolyte (**electrolysis**).

With molten compounds such as sodium chloride the changes are simple:

$$2Na^+Cl^-(l) \rightarrow$$
$$2Na(l) + Cl_2(g)$$

The elements are produced.

With ionic solutions, the processes are more complex. The products depend on the **concentration** of the solution, the **voltage** which is applied, and the type of **electrode** which is used.

Example:

Electrolyte	Anode	Reaction
$CuSO_4(aq)$	(a) carbon or platinum	$4OH^-(aq) \rightarrow$ $2H_2O(l) + O_2(g) + 4e^-$
	(b) copper	$Cu(s) \rightarrow Cu^{2+}(aq) + 2e^-$

In one case (a) oxygen is released; in the other (b) the copper anode dissolves. Process (b) is used in the purification of copper.

Electron Electrons are very small particles with an extremely tiny mass (about $1/1840$ of that of the **proton**). They have a negative charge and move round the nuclei of atoms. They are given the symbol e^-.

When atoms lose or gain electrons they form ions:

gain $\rightarrow$ negative **anions**, e.g. Cl^- OH^- Br^- S^{2-}
lose $\rightarrow$ positive **cations**, e.g. Na^+ H^+ Fe^{2+} Al^{3+}

An electric **current** is electrons moving through a conductor. The way in which electrons are located in an atom is known as the **electronic configuration**.

Electronegativity The electronegativity of an

atom is the power that it has to attract **electrons** to itself in a chemical bond.

Li   Be   B   C   N   O   F

Na   electronegativity increases *across* a   Cl
K   period and *up* a group   Br
Rb   I
Cs   At

It follows that **fluorine** has the *most* electronegative atoms.

Electronic configuration (arrangement)

Electrons move round the **nucleus** of an **atom**. The movement takes place in clearly-defined regions (orbitals) in the atoms. Electrons are pictured as being arranged in **shells** each of which can contain a *maximum* number of electrons.

Shell number:	①	②	③	④	⑤
Maximum number of electrons:	2	8	18	32	50

As we go across the **periodic table** the number of electrons increases by one per atom. The shells are filled up in turn – ① first, then ②, etc. – so that the electrons go into the region of the atom which has the lowest **energy**. For some elements described in this book, the electronic configuration is given, e.g. copper: $2 \cdot 8 \cdot 18 \cdot 1$, i.e. the ①, ② and ③ shells are full and the ④ shell is beginning to fill.

Different configuration patterns are found in the periodic table, e.g. **group** I elements always have an electronic configuration ending in 1:

| Lithium | 2·1 |
| Sodium | 2·8·1 |

The **halogens** have a different pattern:

Fluorine	2·7
Chlorine	2·8·7
Bromine	2·8·18·7

Electroplating If a metallic object is made the **cathode** in an **electrolysis** where the **electrolyte** contains metal **ions**, the object can be coated (plated) with a very thin layer of metal. Conditions – **current**, electrolyte **concentration** and **temperature** must be carefully controlled.

Electroplating is usually carried out for either (a) protection, e.g. chromium plating on handle-bars and motor car bumpers and tin plating on cans, or (b) decoration, e.g. silver plating cutlery and ornaments, and gold plating jewellery, etc.

Electrovalent bond See **Ionic bond**.

Element A **pure** substance which cannot be broken down into anything simpler by chemical means is said to be an element. There are 104 elements known to man. Most of these occur naturally on the earth but several have been made in laboratories by **nuclear reactions**. All elements have a unique number of **protons** in their atoms. There are three classes of element:

metals	non-metals	metalloids
e.g. **iron**	e.g. **oxygen**	e.g. **arsenic**

Empirical formula The empirical formula of a compound shows the atoms that are present in the simplest ratio:

Compound	Molecular formula	Empirical formula
Ethene	C_2H_4	CH_2
Butane	C_4H_{10}	C_2H_5
Propane	C_3H_8	C_3H_8*
Ethanoic acid	$C_2H_4O_2$	CH_2O

* In some examples the empirical and molecular formulae are the same.

See **Molecular formula** and **Percentage composition**.

Endothermic reaction This is a reaction in which **heat energy** is taken in from the surroundings when a reaction occurs.

There is either a fall in temperature when the reaction occurs, e.g. when sodium nitrate dissolves in water, e.g:

$$NaNO_3(s) \rightarrow NaNO_3(aq)$$

or heat energy has to be continually supplied to make a reaction occur, e.g:

$$CaCO_3(s) \rightarrow CaO(s) + CO_2(g)$$

In endothermic reactions, there is *more* **energy** in the bonds of the products than there was in the bonds of the reactants.

Energy   Energy can be defined as the 'ability to do useful work'. Energy is *locked* inside the nuclei of atoms. Sometimes this can be released – see **Nuclear reactions**. Energy is also found in the **bonds** between atoms. When chemicals react bonds break and new ones form. In this breaking and forming energy can be *absorbed* or *released* as heat energy – see **Energy change**. But in batteries it is released as electrical energy. Energy is usually measured in **joules** or **kilojoules**. See **Enthalpy**.

Energy change   When chemical reactions occur, heat energy can be released from the reactants, – **exothermic**; or can be taken in by the reactants, – **endothermic**.

Energy changes are measured in **joules** or **kilojoules** and refer to a particular amount of substance – usually a **mole**. So, the energy change for the **Haber process** is −92 kilojoules per **mole** (−92 kJ/mol).

End-point   This is the point at which a reaction is complete.

The end-point of a **titration** is when *all* of one of

the reactants has been used up. The end-point can be identified by an **indicator** or an instrument such as a **pH** meter or a conductivity meter.

Enthalpy *(H)* This is another word for the amount of **heat energy** possessed by a chemical substance. It is given the symbol H. It cannot be measured but, when chemicals react the difference in the heat energy between the reactants and products can be measured by using a **thermometer**. This difference is the enthalpy change and is given the symbol ΔH.

Enzyme Enzymes are usually **protein** molecules. They are **catalysts** which are found in living tissue. They allow complicated biochemical reactions to occur at low temperatures and pressures inside the body. It is thought that an enzyme is a particular shape and this shape allows two reacting molecules to come close together on the enzyme. This makes the reaction easy to carry out, e.g:

$$\text{starch} \xrightarrow{\text{ptyalin}} \text{maltose}$$

reactants enzyme

the reaction the product

Equation In chemistry, a reaction can be described by an equation. The equation can be a *word* equation:

$$\text{hydrogen} + \text{oxygen} \rightarrow \text{water}$$

or it can be a *formula* equation:

$$H_2 + O_2 \rightarrow H_2O$$

An equation is said to be *balanced* if there are the same number of each kind of atom on each side of the equation, e.g:

$$2H_2 + O_2 \rightarrow 2H_2O$$

is a balanced equation.

This tells us that two **moles** of hydrogen react with one mole of oxygen to produce two moles of water. It also tells us that 4 g of hydrogen react with 32 g of oxygen to produce 36 g of water:

$$2H_2 + O_2 \rightarrow 2H_2O$$
$$2(2) + 32 = 2(2 + 16) = 36$$

Sometimes we use **ionic equations**:

$$Cu^{2+} + Zn \rightarrow Cu + Zn^{2+}$$

State symbols are usually added after the formula to denote which **state of matter** the substance is in, e.g. gaseous hydrogen and oxygen react to produce liquid water:

$$2H_2(g) + O_2(g) \rightarrow 2H_2O(l)$$

Equilibrium In a **reversible reaction** the reaction proceeds in both directions, e.g:

$$3Fe(s) + 4H_2O(g) \rightleftharpoons Fe_3O_4(s) + 4H_2(g)$$

The reactants (iron and steam) produce the oxide and hydrogen. As soon as the products are made, they begin to react to reform steam and iron. There eventually comes a time when the **rate** of the forward $\rightarrow$ reaction equals the rate of the reverse $\leftarrow$ reaction. At this point the proportion of substances present is constant. There seems to be no reaction taking place. There is an *equilibrium* between the forward and reverse reactions.

It is important to realize that such an equilibrium is dynamic – the reactions are occurring but because they are taking place at the same rate no charge can be seen.

Many important reactions involve equilibria.

In the **Haber process**, with the usual operating conditions *less than 20%* of the **hydrogen** and **nitrogen** are converted into **ammonia**. This is the equilibrium **yield** of ammonia in the reaction.

Ester An ester is a compound formed between an **alcohol** and a carboxylic **acid**.

A **strong acid catalyst** is usually needed for the reaction.

Esters often have sweet, *fruity* smells and are used in perfumes, flavourings and essences.

ethanoic   ethanol   ethyl ethanoate
acid   (CH$_3$ COOC$_2$ H$_5$)

The reaction between acid and alcohol is known as *esterification*.

Ethane Ethane is an **alkane**. It is a colourless, flammable gas which is found in **natural gas** to a small extent.

C$_2$H$_6$

Ethanoic acid (acetic acid) Ethanoic acid is a colourless liquid (**melting point** 16°C, **boiling point** 118°C).

It is a carboxylic acid which is made industrially

by the oxidation of **naphtha**. It is used in the production of important fabrics such as *Tricel*.

Ethanoic acid is present in **vinegar** (about a 4% aqueous solution).

Although it is a **weak acid**, in a concentrated form it can cause severe skin burns.

$CH_3 CO_2H$

Ethanol Ethanol is a colourless, flammable **alcohol** whose **boiling point** is 78°C. Ethanol is the alcohol contained in alcoholic drinks. These are made by **fermentation**, but in the chemical industry it is made by the hydration of **ethene**:

$$C_2H_4(g) + H_2O(l) \rightarrow C_2H_5OH(l)$$

C_2H_5OH

Ethanol is used in industry as a solvent. It is also used in **methylated spirits**. It can be oxidized to ethanol and **ethanoic acid**.

Ethene (ethylene) Ethene is a gaseous **alkene** which is produced by **cracking alkanes** (e.g. **naphtha**). Because it is **an unsaturated** compound it is reactive. It is a very important chemical being used to make **plastics** such as **poly(ethene)**.

C_2H_4

Ethene can **add** molecules across its **double bond**, e.g. to produce **ethanol**:

C_2H_5OH

$$C_2H_4Br_2$$

+
Br—Br

These are addition reactions.

Ethers Ethers are **organic compounds** which have an oxygen atom which is bonded to two carbon atoms.

$$C_4H_{10}O$$

One of the most important is diethyl ether (ethoxyethane). Diethyl ether is usually known simply as *ether*. This is the ether which was once widely used as an anaesthetic and which is today used as a solvent for substances which do not dissolve in water. It is very flammable and air/ether mixtures are dangerously explosive.

It is produced by the action of concentrated sulphuric acid on **ethanol**.

Ethylene This is the name by which **ethene** was originally known.

Ethyl ethanoate   This is the **ester** produced by reacting **ethanol** and **ethanoic acid**.

Ethyne (acetylene)

Ethyne is a gaseous **alkyne** which is flammable and which forms explosive mixtures with **air**. It can be produced by the **cracking** of petroleum products, although it used to be made by the action of water on calcium carbide:

$$CaC_2(s) + 2H_2O(l) \rightarrow C_2H_2(g) + Ca(OH)_2(aq)$$

The gas was once widely used for illumination, being burnt in special lamps. They were used in mines and on bicycles. Today ethyne is used to make **chloroethene** and thence **poly(chloroethene)**. It is also used to produce very high **temperatures** by being burnt in the oxy-acetylene torch. Temperatures of 2700 °C can be obtained.

Evaporation   This is a change of state from **liquid** to **vapour** which can occur at any temperature up to the boiling point. It takes place because molecules escape from the body of the liquid into the **atmosphere**. Only a small proportion of the molecules have sufficient energy to escape at any time but over a period they will all escape. Generally speaking, the *lower* the boiling point, the faster will be the rate of evaporation.

some molecules have enough energy to escape

molecules moving at different speeds in the liquid.

Exothermic reaction In an exothermic reaction heat energy is released to the surroundings from the reactants. The **bonds** of the products contain less energy than the bonds of the reactants. The products are *more* stable than the reactants.

Many common reactions are exothermic. All **combustion** and **neutralization** reactions are exothermic:

The **Haber process**

$$3H_2(g) + N_2(g) \rightarrow 2NH_3(g)$$

and **Contact process**

$$2SO_2(g) + O_2(g) \rightarrow 2SO_3(g)$$

are both exothermic reactions. **Enthalpy** changes (ΔH) are less than zero (negative).

See **Endothermic reactions** and **Energy changes**.

95

Explosion An explosion is a rapid expansion of **gas** which causes a shock wave. Common explosives are gunpowder, nitroglycerine and TNT. These materials react with oxygen if *detonated* and large temperatures are produced. A lot of gas is also produced and the gas molecules and the shock wave produced by the reaction travel outwards from the explosion with great energy. It is this energy which causes the damage in explosions.

In nuclear explosions the vast amount of energy locked within the **nucleus** of atoms is released.

Faraday A Faraday is the amount of electric charge possessed by a **mole** of **electrons** (about 96500 **coulombs**). This amount of electricity is needed to liberate one mole of *atoms* of a *univalent element* (e.g. silver) in an electrolysis:

$$Ag^+ \quad + \quad e^- \quad \rightarrow Ag$$

| One mole of silver ions. | One mole of electrons. | One mole of silver atoms. |

Fehling's test This is a test which is used to detect certain **organic reducing agents**. Aldehydes can be distinguished from ketones, and sugars such as **glucose** from **starch** by using the test. Fehling's solution is a mixture of copper(II) sulphate, sodium hydroxide and sodium potassium tartrate. When it is heated with an appropriate reducing agent, e.g. glucose or ethanol, the copper(II) salt is reduced to a red precipitate of copper(I) oxide (Cu_2O).

Fermentation   This is a process where by **chemical changes** are made to **organic** chemicals by the use of living organisms such as **yeasts** and bacteria. The changes are brought about by **enzymes** acting as **catalysts**.

One important example is the changing of **sugars** to **ethanol** by the action of yeast:

$$C_6H_{12}O_6 \xrightarrow{\text{zymase}} 2C_2H_5OH + 2CO_2$$

glucose   ethanol   carbon dioxide

This reaction is used to make alcoholic drinks:

malted barley ⟶ beer

⟶ whisky (after **fractional distillation**)

fruit, e.g. grapes ⟶ wine

⟶ sherry and port

⟶ brandy (after fractional distillation)

Fertilizer   A substance which is added to the soil to replace the nutrients which have been removed by plants. Fertilizers may be natural, e.g. manure and compost, or artificial, e.g. ammonium nitrate, superphosphate. Compounds containing **nitrogen, phosphorus** and **potassium** (N,P,K) are the main constituents of artificial fertilizers. The addition of fertilizers to the land is an important part of the **nitrogen cycle**.

Filter An insoluble **solid** and a **liquid** can be separated by pouring the mixture into a filter paper in a filter funnel. The liquid (or solution) passes through the paper. This is known as the *filtrate*. The solid (known as the *residue*) is left in the filter paper. The whole process is known as *filtration*.

Fission This means *breaking up*. If a molecule is split into two or more parts it is said to undergo fission. In **nuclear reactions** the fission of a large atom, e.g. **uranium** into two smaller ones, e.g. **barium** and **krypton** releases enormous amounts of **energy**.

Fixing nitrogen This is the term which describes any process which converts atmospheric **nitrogen** into compounds which are useful as fertilizers. Fixing occurs in the **Haber process** and is carried out by bacteria found on the roots of certain plants, e.g. peas and beans.

Flame When substances react and hot **gases** are produced the gases often give off light and **heat energy**. This is known as a flame. Flames vary in their colour and their temperature. The gas burner flame when the air hole is *closed* is yellow because the flame produced contains tiny glowing particles of unburnt carbon. The flame when the air hole is *open* (shown in the diagram) has different parts to it. The temperature varies in the different parts of the flame. An X marks the hottest place.

The burner flame

— outer cone

— inner cone

— unburnt gas

— burner barrel

Flame test This test can be used to identify metal ions present in compounds.

A clean nichrome or **platinum** wire is dipped into concentrated **hydrochloric acid** and then into a sample of the compound under test. The wire which now has a small amount of the compound attached to it is then put into the hottest part of the flame. Metals are characterized by specific colours:

Copper	green/blue	Potassium	lilac
Calcium	brick red	Barium	apple green
Sodium	yellow/orange	Lithium	bright red

Flask Flasks are containers (usually glass) in which chemical reactions take place. Some have graduations on them to show specific **volumes**.

flat bottomed   round bottomed   retort   conical

Fluorine (F_2) Fluorine is a gaseous, nonmetallic element in group(VII) of the periodic table. It is a **halogen** and the most reactive element known.

Fluorine is a vigorous oxidizing agent which will even oxidize chloride ion and water:

$$2Cl^-(aq) + F_2(g) \rightarrow 2F^-(aq) + Cl_2(aq)$$
$$2H_2O(l) + F_2(g) \rightarrow 4HF(g) + O_2(g)$$

Compounds of carbon and fluorine (fluorocarbons) are important refrigerants and aerosol propellants.

Fluorine is extracted by the **electrolysis** of a molten fluoride, e.g. KHF_2 and fluoride compounds are added to toothpaste to reduce tooth decay. See **Poly(tetrafluoroethene)** (PTFE).

Formalin This is an **aqueous solution** of methanal (formaldehyde).

Formula See **Empirical formula** and **Molecular formula**.

Fractional distillation (fractionation) A **solvent** can be separated from a **solution** by the process of simple **distillation**. For a mixture of liquids with different **boiling points** it is usual to distil it and collect the products (*distillates*) which boil in definite temperature ranges. These distillates are termed *fractions*. The diagram below shows how **petroleum** might be fractionally distilled.

The exact temperature ranges of these fractions can be varied depending on the product which is in greatest demand. The distillation of whisky and brandy also involve fractional distillation.

102

Fructose This is a sugar molecule. Its formula is $C_6H_{12}O_6$ and it is found in Golden Syrup and honey.

Fuel A fuel is a substance which releases **heat energy** when it is treated in a certain way. In most fuels, the energy is released by combustion, e.g:

Fossil fuels		Organic fuels
Natural gas	⎫ Burn in	⎰ Wood
Petroleum	⎬ oxygen to	⎱ Waste materials
Coal	release	
Peat	carbon	
Coke	dioxide and	
	water.	

Nuclear fuels, e.g. **uranium** and **plutonium** produce heat because of changes that occur inside the atom. These **nuclear reactions** generate very large amounts of energy.

Fusion This means *coming together* and has two particular uses in chemistry.
  1) Fusion is another term for melting. See **Latent heat**.
  2) Some **nuclear reactions** involve two atoms coming together to form a single atom. This is the opposite of **fission** and can involve the release of enormous quantities of energy. Hydrogen bombs are fusion weapons. Attempts are being made to use fusion reactions to generate electricity.

(g) This is a *state symbol* and is used to denote that a substance is a **gas**:

$CO_2(g)$, $\quad O_2(g)$, $\quad NH_3(g)$, $\quad H_2O(g)$, $\quad H_2(g)$.

Galvanizing Iron and steel sheeting is often coated with a thin layer of zinc. This is called *galvanized iron*. Zinc is more resistant to corrosion than iron and so it can protect the metal. Also, if the coating is scratched and the iron and zinc come into contact with a liquid and a **battery** is set up, the zinc reacts rather than the iron – it is more reactive. So, galvanizing protects the iron even when the protective layer is broken.

Gamma rays A form of high energy electromagnetic radiation. They are produced in **nuclear reactions** and have great penetrating powers. They are similar to **X-rays**. They are used in the treatment of cancer – they kill body cells. They are also used to sterilize substances.

Gas A **state of matter** in which atoms and molecules have few bonds between them and consequently have a large amount of freedom of movement.

When heat energy is supplied to a liquid, the atoms or molecules are given enough **kinetic energy** to overcome the bonds which hold them together in the **liquid** state. In a gas, atoms or molecules fill the whole container and their movement against the walls of the container exerts a **pressure**.

Gas laws If **Boyle's law** and **Charles' law** are combined we realize that for a *fixed mass of gas*:

$$\frac{(Pressure) \times (Volume)}{(Temperature)} = constant$$

or: $\dfrac{PV}{T} = constant$

This means that if we compare gases under different conditions, e.g. different **temperature** and **pressure** then:

$$\frac{P_1 V_1}{T_1} = \frac{P_2 V_2}{T_2} = constant$$

condition 1   condition 2

This allows us to obtain much useful information about gases by calculation rather than by measurement. See STP.

Giant structure Molecules are made up of small numbers of atoms bonded together. In giant structures of **atoms** or **ions** there are large numbers of particles present in a **crystal lattice**. Each particle has a strong force of attraction for all the other particles which are near to it. In this way attractive forces are spread through the structure and giant structures tend to have high **melting** and **boiling**

points. Ionic substances are giant structures and most elements, e.g. all metals and several non-metals.

Glass Glass is a hard transparent mixture of silicates. The cheapest and commonest kind of glass is called *soda glass*. It is made by heating together sand, sodium carbonate, calcium oxide and broken glass (cullet). The hot liquid is cooled down very slowly. The slowness of cooling means that the glass does not crystallize. Glass is a *supercooled liquid*.

In *lead glass*, the sodium carbonate is replaced by lead(II) oxide. This gives a glass with a high refractive index. It is used in making crystal glassware. See **Borosilicate glass**.

Glucose Glucose is a **monosaccharide** molecule. It is found in honey, Golden Syrup and fruits. All **sugar** and **starch** which enters our bodies is converted to glucose. It is then used to provide **energy**.

Glycol This material consists of ethane-1, 2-diol and is used as **antifreeze** material in engines. The compound contains two alcohol (OH) groups and is made from **ethene** by oxidation to epoxyethane and then water is added to give the diol. Ethene is obtained by the **cracking** of **petroleum**.

ethene   epoxyethane   ethane-1, 2-diol

Gold

Gold is a valuable metal which is prized for its use as jewellery. It is used as a substitute for money. It is found uncombined with other elements and the major deposits are in South Africa and the USSR. Chemically it is very *inert* reacting only with vigorous **oxidizing agents** such as **chlorine** and certain acids, e.g. *aqua regia*. Gold is a soft metal and for most uses it is alloyed with copper or silver.

Pure gold is said to be twenty-four carat gold. Nine carat gold is nine parts gold to fifteen parts copper, i.e. 37.5% gold. This is hard and is commonly used in jewellery.

Gramme (g)   This is a unit of **mass**. It is $^{1}/_{1000}$ of a kilogram and is used in all scientific work. The symbol (g) is used, e.g. 100 g.

Graphite This is an allotropic form of **carbon**. It is found naturally as *plumbago*. *Charcoal* consists of small particles of graphite. It is a **giant structure** with the carbon atoms bonded together in a hexagonal shape. There are strong **bonds** between the atoms in the planes but the bonds between the atoms in different planes are weak. It is possible for the planes to slip over each other.

carbon atoms in hexagonal plates

strong bonds

weak bonds

the plates slip

Because of this property, graphite is very useful as a lubricant and, of course, in pencils where layers of graphite are left behind on the paper.

Graphite also finds use as electrodes in the extraction of elements, e.g. sodium, aluminium and chlorine.

108

Group In the **periodic table** the **elements** are arranged in horizontal **periods** and vertical *groups*. In these groups the elements have similar chemical properties, e.g.

Group I	lithium, sodium, potassium
Group II	beryllium, magnesium, calcium
Group III	boron, aluminium
Group IV	carbon, silicon
Group V	nitrogen, phosphorus
Group VI	oxygen, sulphur
Group VII	fluorine, chlorine, bromine
Group 0	helium, neon, argon

Each group has a characteristic **electronic configuration**.

Haber process Ammonia is made in this process from **hydrogen** and **nitrogen**. Nitrogen is obtained from the **air** and hydrogen from the steam **reforming** of **natural gas**. The gases are reacted together in a 1:3 ratio at 500°C and two hundred **atmospheres pressure**. An **iron catalyst** is used:

$$N_2(g) + 3H_2(g) \rightleftharpoons 2NH_3(g)$$

It is an **equilibrium reaction** and the conditions are chosen to produce as much ammonia as possible in the shortest time. Under these conditions about 15% of the reactants are converted to ammonia. Using a *lower* temperature will generate *more* ammonia but at a much slower **rate**. Conversely, using a *higher* temperature will produce the ammonia more quickly, but the **yield** will be lower. The chosen conditions are the *optimum* ones.

Haemoglobin This is the red pigment which is found in red blood corpuscles. It contains an atom of iron in the complex molecule. **Oxygen** reacts with the pigment forming *oxyhaemoglobin* and oxygen is released from the complex where it is needed. **Carbon monoxide** also reacts with haemoglobin, forming *carboxyhaemoglobin*. This is a very stable compound and prevents the carriage of oxygen. This is why carbon monoxide is such a dangerous poison. It is important that the body obtains sufficient iron each day to maintain the correct levels of haemoglobin. The daily requirement is about 11 mg. Insufficiency leads to the condition known as *anaemia*.

Half-life When a radioactive **isotope** gives off **alpha** or **beta particles** (decays) it changes into a different isotope. As the decay occurs the number of nuclei becomes fewer. The *half-life* of an isotope is the time taken for the **radioactivity** to decrease to half of its original value. The decay of an isotope is usually traced by measuring the **rate** at which particles are emitted. This rate is proportional to the number of nuclei present.

The plot shows the decay of an isotope with a half-life of one minute, i.e. in each minute the number of particles emitted falls by half e.g. 4000–2000, 2000–1000, etc.

There is great variety in the length of half-lives from isotope to isotope. The half-life of each isotope is constant, however, under all conditions of **temperature** and **pressure**.

isotope	half-life	
Carbon-14	5730	years
Oxygen-20	14	seconds
Copper-64	756	minutes
Uranium-234	250,000	years

Half-reaction It is often useful to represent a reaction by two half reactions, e.g. the **displacement** of **copper ions** from solution by zinc:

$$Cu^{2+}(aq) + Zn(s) \rightarrow Zn^{2+}(aq) + Cu(s)$$

can be viewed as

i) $Cu^{2+}(aq) + 2e^- \rightarrow Cu(s)$

ii) $Zn(s) \rightarrow Zn^{2+}(aq) + 2e^-$

From these half-reactions it is easy to see that the zinc is being oxidized by the copper ions and that the copper ions are being reduced by the zinc. Half-reactions are particularly useful for looking at **redox** reactions.

Halide A halide is a compound of a **halogen** with another **element**.

Examples:

hydrogen chloride HCl sodium bromide NaBr

calcium fluoride CaF_2 silicon tetrachloride $SiCl_4$

Metal halides tend to be **ionic**, whereas non-metal halides have **covalent bonds**.

Metal halides can be produced by reacting the elements together:

$$2Na(s) + Cl_2(g) \rightarrow 2NaCl(s)$$

Halogens Elements in **group** VII of the **periodic table** are called halogens. They are all poisonous, non-metallic elements. **Fluorine** and **chlorine** are gases, **bromine** is a liquid and **iodine** is a solid at room temperature. They are all **oxidizing agents**. Their oxidizing power and chemical reactivity decreases in the order:

$$F_2 > Cl_2 > Br_2 > I_2 \text{ i.e. as the group is descended.}$$

They react vigorously with **metals** and **hydrogen** forming **halides**. They all contain seven **electrons** in the outer **shell** of the atom and form **anions**, e.g. $Cl^- Br^-$.

Hardness of water Water from the mains supply is described as being *soft* or *hard*. Hard water has passed over and through rocks which have dissolved in the water, e.g. **limestone, chalk** and gypsum ($CaSO_4$). Soft water has collected in areas of the country where the rocks are insoluble in water, e.g. the granite areas such as the Lake District, Cornwall and the Highlands of Scotland.

There are two types of hard water:

Permanent Contains Ca^{2+} and SO_4^{2-} ions. Can be removed by adding sodium carbonate which precipitates the calcium ions as the carbonate.

Temporary Contains Ca^{2+}, Mg^{2+} and HCO_3^- ions. Can be removed by boiling to form the *insoluble* carbonates.

Soap forms a scum with hard water because of the formation of insoluble calcium salts. Soapless **detergents** do not do this. The use of kettles and immersion heaters with hard water becomes increasingly difficult and expensive because **calcium carbonate** is deposited on the heating elements making them inefficient. Substances such as *calgon* and *permutit* are used to make water softer through the use of **ion exchange**.

Heat energy All substances possess heat energy. The higher their **temperature**, the more energy they possess. The source of this energy is the movement of atoms and molecules, i.e. their **kinetic energy**.

Heat energy is taken in during **endothermic reactions** and given out during **exothermic reactions**.

Helium (He)

Helium is an **inert gas**. It is found in **natural gas** and is present in the **atmosphere** to a very small extent. It is completely unreactive and the gas is **monatomic**. It is used in airships as it is eight times less dense than air. It is also used in breathing apparatus for deep-sea divers.

Heterogeneous reaction When chemical reactions occur between substances which are in different physical states, i.e. gases, liquids and/or solids, the reactions are said to be heterogeneous. Hetero ≡ different.

See **Homogeneous reaction**.

Examples:

$$Zn(s) + 2HCl(aq) \rightarrow ZnCl_2(aq) + H_2(g)$$
$$H_2O(l) + Na(s) \rightarrow NaOH(aq) + H_2(g)$$
$$Na(s) + Cl_2(g) \rightarrow 2NaCl(s)$$
$$CuCO_3(s) \rightarrow CuO(s) + CO_2(g)$$

Hofmann voltameter The voltameter is filled by opening the two taps and pouring the electrolyte into the central tube until the apparatus is full. When the taps are closed and the **current** switched on any gases which are produced will collect in the outside tubes.

The **volume** of the gases can be measured and the gases collected.

Homogeneous reaction Reactions are said to be homogeneous when all the reactants and products are in the same physical state. Homo ≡ same.

Examples:

$$Fe(s) + S(s) \rightarrow FeS(s)$$
$$2NH_3(g) + O_2(g) \rightarrow N_2(g) + 3H_2O(g)$$

Homologous series A series of compounds with the same *general* formula, e.g.

alkanes C_nH_{2n+2}
alkenes C_nH_{2n}

The compounds have the same function group(s) and have the same chemical **properties**. The physical properties of the series of compounds change only gradually.

Examples:
Alkanes

methane CH_4
$n = 1$

ethane C_2H_6
$n = 2$

$$H-\overset{\overset{\displaystyle H}{|}}{\underset{\underset{\displaystyle H}{|}}{C}}-\overset{\overset{\displaystyle H}{|}}{\underset{\underset{\displaystyle H}{|}}{C}}-\overset{\overset{\displaystyle H}{|}}{\underset{\underset{\displaystyle H}{|}}{C}}-H$$

propane C_3H_8
n = 3

Hydrate A hydrate is a compound which has **water** chemically combined within it. *Hydrated* salts, i.e. those which contain **water of crystallization** are the best examples.

$$CuSO_4.7H_2O \qquad CaCl_2.6H_2O$$

Hydride A hydride is a compound which contains **hydrogen** and another **element** only. Examples:

NH_3 ammonia H_2O water H_2S hydrogen sulphide

The hydrides of non-metals are **covalent** compounds. Some metal hydrides contain the **ion** H^-, e.g. Na^+H^-. These compounds are very reactive and will readily decompose water. They are used as **dehydrating agents**.

Hydrocarbon A compound of **hydrogen** and **carbon** only is a hydrocarbon, e.g. **alkanes, alkenes, alkynes**:

methane	CH_4	ethyne	C_2H_2
ethene	C_2H_4	benzene	C_6H_6

Hydrochloric acid (HCl(aq)) This is a solution of **hydrogen chloride** in **water** and it contains **chloride** and **oxonium** ions. The maximum **concentration** of the solution is 36% (about 11 **mol/**

117

dm^3). The acid is **monobasic** and produces **salts** called **chlorides**, e.g.

$$Fe(s) + 2HCl(aq) \rightarrow FeCl_2(aq) + H_2(g)$$
$$Mg(s) + 2HCl(aq) \rightarrow MgCl_2(aq) + H_2(g)$$

It is a strong acid being fully dissociated into $Cl^-(aq)$ and $H_3O^+(aq)$ ions in dilute solution. It releases **carbon dioxide** from **carbonates** and **hydrogencarbonates** and can be oxidized to **chlorine**:

$$MnO_2(s) + 4HCl(aq) \rightarrow$$
$$MnCl_2(aq) + 2H_2O(l) + Cl_2(g)$$

Hydrogen (H_2)   Hydrogen is a gaseous **diatomic element**. The atom consists of one **proton** and one electron. The **isotope deuterium** contains a **neutron** in the nucleus of the atom. Hydrogen is very reactive it can form **covalent bonds** by sharing electrons, e.g.

$$2H_2(g) + O_2(g) \rightarrow 2H_2O(g)$$
$$C_2H_4(g) + H_2(g) \rightarrow C_2H_6(g) \ (hydrogenation)$$

118

Two hydrogen ions can be produced by losing the electron to form H^+ or by gaining an electron to form H^- (present in some **hydrides**). The positive hydrogen ion (H^+) is such a small, reactive species (it is a lone proton) that it does not exist in solution. In **aqueous** solution it reacts with the water to form the **oxonium ion. Acids** contain the oxonium ion.

Hydrogen is a **reducing agent**. It is used to convert vegetable oils into *margarine*. Large quantities are used in the **Haber process**. Industrially, the gas is made from **petroleum** by the steam reformation of **alkanes**.

Hydrogen can be used to reduce some oxide and sulphide ores to the metal.

In the laboratory, the gas is made by reacting a metal with an acid other than nitric acid, e.g:

$$Zn(s) + 2HCl(aq) \rightarrow ZnCl_2(aq) + H_2(g)$$

It is also released by the electrolysis of aqueous solutions which contain ions of elements above hydrogen in the electrochemical series, e.g. $NaCl(aq)$, $Mg(NO_3)_2(aq)$, and by the reaction of water with alkali and alkaline earth metals, e.g. sodium, calcium.

Hydrogen is a flammable gas which causes explosive mixtures with oxygen. Great care must be taken in its preparation, collection and use.

Hydrogen bond A weak chemical bond can be formed between **hydrogen** atoms and atoms of **oxygen** or **nitrogen**. Although weaker than **covalent** or **ionic bonds**, hydrogen bonds affect the physical properties of compounds. Molecules in **water** and **ice** are extensively hydrogen bonded to each other.

hydrogen
bonds

Hydrogenation In this process an **unsaturated** compound is turned into a **saturated** one by the addition of **hydrogen**. A **catalyst** such as nickel is used, e.g:

120

ethene   ethane

Hydrogencarbonates (HCO_3^-)

These are the **acid salts** of *carbonic acid* H_2CO_3. The best known examples are:

Sodium hydrogencarbonate $NaHCO_3$. This is the familiar household chemical *bicarbonate of soda*. It is found in self-raising flour and **baking powder**.

Calcium hydrogencarbonate $Ca(HCO_3)_2$. This is found in hard water because rain water passes over carbonate rocks such as limestone:

$$CaCO_3(s) + CO_2(aq) + H_2O(l) \rightarrow Ca(HCO_3)_2(aq)$$

Hydrogen chloride (HCl)

Hydrogen chloride is a colourless gas which is very soluble in water. An aqueous solution is called **hydrochloric acid**. The gas can be made by reacting the elements together or by treating **sodium chloride** with concentrated **sulphuric acid**:

$$H_2(g) + Cl_2(g) \rightarrow 2HCl(g)$$
$$NaCl(s) + H_2SO_4(l) \rightarrow NaHSO_4(s) + HCl(g)$$

The gas will react with ammonia to form dense white fumes of **ammonium chloride**:

$$NH_3(g) + HCl(g) \rightarrow NH_4Cl(s)$$

The bonding in hydrogen chloride is **covalent** but **ions** are formed when it dissolves in a **polar** solvent, e.g. water.

Hydrogen halides These are gaseous compounds formed between **hydrogen** and the **halogens**. They are **covalent** compounds which readily dissolve in water to form an acidic solution which contains the **halide ion**, e.g:

$$HBr(g) + H_2O(l) \rightarrow H_3O^+(aq) + Br^-(aq)$$

Hydrogen fluoride	HF
Hydrogen chloride	HCl
Hydrogen bromide	HBr
Hydrogen iodide	HI

Hydrogen ion The positively charged **hydrogen ion** is a **proton**:

$$H \rightarrow H^+ + e^-$$

This is such a small particle that it is very reactive. In solution, it is chemically combined with the solvent molecules. In water this is represented by the **oxonium ion**:

$$H^+ + H_2O \rightarrow H_3O^+ \left[\begin{array}{c} H \diagdown O \diagup H \\ | \\ H \end{array} \right]^+ \text{ oxonium ion}$$

The concentration of hydrogen ions are expressed in **pH** units, See **Hydroxide ion**.

Hydrogen peroxide (H_2O_2) This compound is usually used as an **aqueous** solution. It can be decomposed to give oxygen:

$$2H_2O_2(aq) \rightarrow 2H_2O(l) + O_2(g)$$

A manganese(IV) oxide catalyst brings about the reaction. It is used as a **disinfectant** and **bleach** in the home and is a powerful **oxidizing agent**.

Hydrogen sulphide (H_2S) This is a colourless gas with a sweetish, sickly odour which reminds you of rotten eggs. The gas is produced when **organic** matter containing **sulphur** rots. It is often found associated with petroleum. It is made in the laboratory by the action of an acid on a metal sulphide, e.g:

$$FeS(s) + 2HCl(aq) \rightarrow FeCl_2(aq) + H_2S(g)$$

The gas is *very* poisonous. Like hydrogen cyanide [HCN] it prevents the transmission of energy within the body. The gas can be detected by the fact that it turns a piece of filter paper soaked in a lead(II) salt to a black colour. This is because insoluble lead(II) sulphide is formed:

$$Pb^{2+}(aq) + H_2S(g) \rightarrow PbS(s) + 2H^+(aq)$$
$$\text{black}$$

A solution of hydrogen sulphide in water is a **weak acid** and the gas is a **reducing agent**.

Hydrolysis This is the term for the decomposition of a substance by the action of water. The water is also decomposed. **Esters** are hydrolysed:

ethyl ethanoate water **ethanoic ethanol
acid**

Hydrophilic This term means *water-loving*. It is used to describe parts of molecules which readily dissolve in water, e.g. the ionic end of a **detergent** molecule is hydrophilic. This *attraction* for water is why detergents work. The other end of the molecule is **hydrophobic** – *water-hating*.

The hydrophobic part of a molecule does not dissolve in water and is not attracted to water. In a detergent this end of the molecule bonds with grease or oils in the material being cleaned. See **Soap** and **Detergent**.

Hydrophobic See **Hydrophilic**.

Hydroxide ion (OH^-) This ion is found in all **alkalis**, e.g. sodium hydroxide, and in alkaline solutions. It is present in *all* aqueous solutions because of the **dissociation** of water:

$$H_2O \rightleftharpoons H^+ + OH^-$$

Solutions with *more* hydroxide ions than hydrogen ions are described as alkaline and have a **pH** greater than seven. Group I hydroxides are soluble in water. Some other hydroxides are sparingly soluble, e.g. $Mg(OH)_2$ $Ca(OH)_2$. Insoluble hydroxides can be precipitated by the use of a soluble hydroxide, e.g:

$$Pb(NO_3)_2(aq) + 2NaOH(aq) \rightarrow$$
$$Pb(OH)_2(s) + 2NaNO_3(aq)$$

Hygroscopic A hygroscopic substance can take in up to 70% of its own mass of water without dissolving or getting wet. Examples are copper(II) oxide, sodium chloride and **silica gel**. See **Deliquescence**.

Ice Ice is solid, crystalline water. Its **melting point** at a **pressure** of one **atmosphere** is 0 °C. The structure of ice is shown here. The regular, crystalline structure, means that frozen water can take on the patterns that we see in snow flakes. Ice is less dense than water and it floats.

hydrogen bonds

Immiscible When two **liquids** do not mix together but form two layers – one liquid on top of the other, they are described as immiscible. One liquid will probably be **polar**, e.g. water, and the other non-polar, e.g. **ether**. Two polar liquids will mix completely, e.g. **water** and **ethanol**, as will two non-polar liquids, e.g. ether and **tetrachloromethane**.

two immiscible liquids

less dense liquid

more dense liquid

a mixture of miscible liquids

Indicator An indicator is a substance which changes its colour under different conditions. It *indicates* the **end-point** of a reaction. Indicators are useful in **acid-base titrations** where they are substances which have different colours at different **pHs**. They are also used in titrations between **metal ions** and **complex ions**.

Acid-base titration	pH range of the end-point	Indicator	colour change acid-base
strong acid – strong base	4–10	phenolphthalein	colourless – red
		litmus	red – blue
strong acid – weak base	4–8	methyl red	red – yellow
		litmus	red – blue
weak acid – strong base	7–12	thymol blue	yellow – blue
weak acid – weak base		No indicator is suitable for these reactions	

Acid-base indicators are **weak acids** and the acid molecule and its negative ion have different colours:

$$\text{Acid molecule} \rightleftharpoons \text{anion} + H^+$$

colour I   colour II

e.g. red   e.g. blue

See **Universal indicator**.

Inert gases These elements are in **Group** o of the **periodic table**. They are all very unreactive **monatomic** gases which occur in the **atmosphere** to small extents:

Helium, Neon, Argon, Krypton and **Xenon**.

Their **electronic configurations** are very stable and the elements show no tendency to lose or gain electrons. Because of this, they find it hard to form compounds. Compounds do exist of the more massive elements, e.g. XeF_4, but these are rare.

Inert gas structure Inert gases are stable, unreactive elements because of their electronic configurations. They have *eight* **electrons** in their *outer* **shell**. This is a stable configuration because the eight electrons completely fill a part of the shell making it difficult to add another or take one away.

When elements form ions, the ions which are formed have the inert gas configuration:

Element	Ion	Electronic configuration		Inert
		atom	ion	gas
Sodium	Na^+	2·8·1	2·8	Neon
Fluorine	F^-	2·7	2·8	Neon
Calcium	Ca^{2+}	2·8·8·2	2·8·8	Argon
Sulphur	S^{2-}	2·8·6	2·8·8	Argon
Aluminium	Al^{3+}	2·8·3	2·8	Neon
Bromine	Br^-	2·8·18·7	2·8·18·8	Krypton

Inorganic chemistry Inorganic chemistry is concerned with non-**organic** aspects of chemistry, i.e. **elements** and their **compounds**. This includes **carbon** the element, its **oxides**, metal **carbonates** and **hydrogencarbonates**, but excludes all **organic** compounds, e.g. **alcohols, esters, ethers, hydrocarbons**, etc.

Insoluble Substances are described as insoluble when they either do not dissolve in a **solvent** or when only a very small amount dissolves.

Insulator An insulator is a substance which is either a poor **conductor** of heat or electricity.

Non-metallic elements are usually insulators, as are most *solid* compounds and **polymers. Graphite** is an exception.

Examples of efficient insulators:

> Expanded polystyrene.
> Rubber.
> Mineral wools, e.g. *Rockwool*.
> Glass fibre.

Iodides Iodides are compounds of **iodine** and another **element**, e.g:

Potassium iodide KI, Hydrogen iodide HI.

Iodine (I_2) Iodine is a shiny, grey non-metal. It is a **halogen** and has a **diatomic** molecular structure.

Iodine is extracted from sodium iodate(v) [$NaIO_3$] and a small amount from seaweed.

Silver iodide is used in photographic emulsions. The human body needs 0.07 mg of iodine per day. Its principal use in the body is in the production of the hormone *thyroxime*. Potassium iodide is often added to table salt to provide this iodine.

Iodine will react with some metals directly to form **iodides**. It will react with **hydrogen** and **chlorine** forming [HI] and [ICl]. The vapour of the element is purple and is choking and **caustic**.

Ion An ion is a species which possesses an electrical charge. **Cations** have a positive charge, **anions** have a negative charge, e.g. sodium Na^+, oxide O^{2-}. Atoms tend to gain or lose **electrons** to produce an ion with an **inert-gas** configuration.

Groups of atoms (radicals) can also possess a charge, e.g:

Sulphate $SO_4{}^{2-}$, Nitrate $NO_3{}^-$,
Hydroxide OH^-, Ammonium $NH_4{}^+$

Ion-exchange It is possible to purify water by passing it over a resin in a tube. With sea water the **sodium ions** would be replaced by **hydrogen** ions and the **chloride** ions by **hydroxide** ions. In other words sodium chloride would be *exchanged* for water.

The ions are bonded to the resin:

and exchange occurs. The product is *de-ionized* water. This method can be used for softening water, e.g. **calcium** ions are replaced by **sodium** ions.

Ionic bonds These are chemical bonds which occur because of the electrostatic attractive forces between negatively and positively charged **ions**.

Ionic bonds occur in compounds of non-metals from groups VI and VII and metallic elements, e.g:

$$Na^+Cl^- \quad Mg^{2+}Br^-$$

and also in compounds involving radicals such as sulphate and nitrate, e.g:

$$Cu^{2+}SO_4^{2-} \text{ and } K^+NO_3^-$$

Ionic compounds have **giant structures**.

Ionic equation The reaction between sodium hydroxide and hydrochloric acid can be described by the molecular **equation**:

$$NaOH(aq) + HCl(aq) \rightarrow NaCl(aq) + H_2O(l)$$

This reaction is an ionic one and can be represented in an *ionic* form:

$$Na^+OH^-(aq) + H^+Cl^-(aq) \rightarrow$$
$$Na^+Cl^-(aq) + H_2O(l)$$

Because Na^+ and Cl^- appear on *both* sides of the equation they do not take any part in the reaction and can be omitted. The *ionic* equation is:

$$OH^-(aq) + H^+(aq) \rightarrow H_2O(l)$$

The other ions are **spectator ions**.

Ionization When an **atom** loses or gains **electrons** it becomes an **ion**. This gain or loss is termed ionization.

Iron Iron is the most widely used metallic element. It is mainly encountered as **steel alloys** and in this form it is used for building girders, machine bodies (cars, cookers, fridges), containers (boxes, cans, drums), tools, utensils and many other items of everyday life. The metal is extracted from ores such as haematite (Fe_2O_3) and magnetite (Fe_3O_4) by reduction with **carbon monoxide** in the **blast furnace**. The iron which is produced in this way is brittle and is made into many different kinds of steel to strengthen it and to give it special properties.

One of the main problems with iron is that it rusts, i.e. it oxidizes in air to produce a soft, crumbly oxide.

Iron $\xrightarrow{\text{(moist air)}}$ Iron(III) oxide (Fe_2O_3)
(strong, useful metal)   (weak, worthless oxide.)

Iron is a **transition metal** and can have **valency** 2 or 3. The metal reacts with dilute acids to form iron(II) compounds, e.g:

$$Fe(s) + 2HCl(aq) \rightarrow FeCl_2(aq) + H_2(g)$$

but will give iron(III) compounds when reacted with vigorous **oxidizing agents** such as **chlorine**:

$$2Fe(s) + 3Cl_2(g) \rightarrow 2FeCl_3(s)$$

Iron has a **reversible reaction** with steam:

$$3Fe(s) + 4H_2O(g) \rightleftharpoons Fe_3O_4(s) + 4H_2(g)$$

Iron(III) compounds can be identified by their giving a red colour with potassium thiocyanate solution (KCNS). Iron(II) compounds do not do this but do produce a blue colour with a solution of potassium hexacyanoferrate(III) $K_3Fe(CN)_6$.

Iron compounds Iron forms compounds with **valency** (II) and (III). Iron(II) compounds tend to be green in colour whilst iron(III) compounds are yellow or brown.

Iron(III) oxide Fe_2O_3 (haematite)	This is used as a yellow/brown pigment in the paper, linoleum and ceramics industries. It is used as a **catalyst** and as a polish, e.g. in jewellers rouge.

Iron(III) chloride $FeCl_3$	This is produced by reacting the elements together. It is an important **catalyst** and also finds use in the purification of water and in the production of pharmaceutical products.
Iron(II) sulphate $FeSO_4.7H_2O$	Iron(II) sulphate can be recovered from the waste materials left behind in the electroplating processes. It is an important compound, used in the preservation of wood, in inks and in *lithography*.

Isomer When two or more *different* compounds have the same **molecular formula** they are described as isomers. Because the compounds are different, they have different properties.

Example:

Formula C_2H_6O

Structure of the compounds

ethanol   methoxymethane

Isotopes Atoms of an **element** always contain the *same* number of **protons** (same **atomic number**). The number of **neutrons** in atoms of an element can be different (different mass number).

A sample of **chlorine** contains atoms which have eighteen neutrons (76% of the total) and atoms which have twenty neutrons (24%). Thus, chlorine is said to have two isotopes:

$$^{35}_{17}Cl \quad ^{37}_{17}Cl$$

The ratio of the isotopes is always constant. Most elements are found in *isotopic forms*. Some examples are shown here:

	Isotopes	
Bromine	$^{79}_{35}Br$ (51%)	$^{81}_{35}Br$ (49%)
Carbon	$^{12}_{6}C$ (99%)	$^{13}_{6}C$ (1%)
Copper	$^{63}_{29}Cu$ (69%)	$^{65}_{29}Cu$ (31%)
Magnesium	$^{24}_{12}Mg$ (79%)	$^{25}_{12}Mg$ (10%)
	$^{26}_{12}Mg$ (11%)	

All the ones quoted are *stable* isotopes. However, many more unstable, radioactive isotopes exist. A few elements have no isotopes, i.e. they have no variation in the number of neutrons that are found in the atom, e.g:

Fluorine
Gold
Iodine

Joule   This is the **SI unit** of **energy** and work. It has the symbol J. In scientific usage the **kilojoule** is common.

Kelvin temperature scale   The Kelvin is the **SI unit** of **temperature**. The Kelvin (K) is the same as one degree **Celcius**. Absolute zero is O K.

Here is a comparison with the Celcius scale:

Kelvin	0	273	310	373	K
Celcius	−273	0	37	100	°C
	absolute zero	freezing point of water	blood heat	boiling point of water	

Kerosine   This is a product of **petroleum** refining. Its uses include fuel for jet aircraft and also household paraffin which can be used for heating and lighting. Kerosines boil between 160°C and 250°C.

Kilo   This is a prefix which means *one thousand*. In common useage, a kilo has come to mean a kilogram, e.g: 'a kilo of potatoes'. This is imprecise and should not be used in scientific work.

Uses:

kilogram	1000 g
kilometre	1000 m
kilojoule	1000 J
kilowatt hour	1000 Wh

Kilojoule (kJ)   One thousand **joules**.

Kinetic energy   Energy a body possesses because of its motion. The greater its velocity (speed) the more energy it has. The kinetic energy of a particle is ½ mv² where m is its **mass** and v its velocity. If the mass is measured in kilograms and the velocity in metres/second the kinetic energy is measured in **joules**.

Kinetic theory   The kinetic theory explains that all particles (**atoms** and **molecules**) are moving and that the extent to which movement can occur depends on the temperature. In **solids** and **liquids** the amount of movement is restricted by bonds between adjacent particles. In **gases** the movement is only restricted by the walls of the container. The higher the temperature, the greater the **kinetic energy** the particles possess. See **Lattice**.

Krypton   Krypton is an **inert gas**. It is found in

the **atmosphere** to a small extent and is used as a filling for electronic valves and fluorescent tubes. Like all inert gases, krypton is **monatomic**.

I This symbol denotes a **liquid**, e.g. $H_2O(l)$, $Hg(l)$, $H_2SO_4(l)$.

L This is the symbol for the **Avogadro constant**. It is the number of particles in one **mole** of a substance. Its value is 6×10^{23} particles per mole.

Latent heat This is the amount of **heat energy** released or absorbed in a change of state at a fixed **temperature** (e.g. **melting point** and **boiling point**). The latent heat of *fusion* is that energy needed to turn one **mole** of **solid** into a **liquid** *at its melting point.*

Similarly, the latent heat of *vaporization* is that energy required to turn one mole of **liquid** into a **gas** at its boiling point.

These symbols can be used:

ΔH_m m = melting ΔH_b b = boiling
(fusion)   (vaporization)

Examples

Water	$\Delta H_m =$	6 kJ/mole
	$\Delta H_b =$	41 kJ/mole
Magnesium	$\Delta H_m =$	9 kJ/mole
	$\Delta H_b =$	129 kJ/mole

Lattice A lattice is a regular arrangement of **molecules, atoms** or **ions** within a crystalline solid. Lattices contain large numbers of particles which are arranged in very particular ways. Small sections of three lattices are shown here. They are a) aluminium b) sodium c) magnesium.

(a) cubic close packed

(b) body centred cubic

(c) hexagonal close packed

Law of constant composition A chemical **compound** always contains the same **elements** combined in the same proportions by **mass**.

Examples:

Water always has hydrogen and oxygen atoms within it in the proportion: $H/O = 2/16 = 1/8$

Carbon dioxide always has carbon and oxygen atoms within it in the proportion: $C/O = 12/32 = 3/8$

Law of multiple proportions If two **elements** form more than one **compound** then there is a simple relationship between the masses of one of the elements and a *fixed* mass of the other element.

Example: Oxides of nitrogen

formula	ratio	simple ratio
	N:O (by mass)	N:O
N_2O	28:16	14:8
NO	14:16	14:16
NO_2	14:32	14:32
N_2O_5	28:80	14:40

There is a simple and steady increase in the ratio of oxygen to nitrogen.

Lead Lead is a dense metallic **element** which is in **group** IV of the **periodic table**. It is a soft, **malleable**, grey element which is extracted from *galena* (PbS). Lead and its **alloys** are widely used, e.g. in the lead-acid **accumulator**, solders, protection from moisture (used on roofs), radioactive isotopes and X-rays. Because of its chemical inertness, lead can be used to store **sulphuric acid**.

Lead compounds Lead forms lead(II) and lead(IV) compounds. Most lead compounds are cumulative **poisons**.

Lead(II) nitrate $Pb(NO_3)_2$	This is the only common **soluble** lead compound.
Lead(II) oxide PbO	This is the yellow oxide, litharge. It is used in making of glass and enamels.
Lead(IV) oxide PbO_2	This oxide is formed in the lead-acid **accumulator**.

Le Chatelier's principle If a reaction is at **equilibrium** and any of the conditions are changed, further reaction will occur to counter the changes and re-establish equilibrium.

Examples:

The **Haber process** $N_2(g) + 3H_2(g) \rightleftharpoons 2NH_3(g)$

If extra **nitrogen** or **hydrogen** is added to an equilibrium mixture then more **ammonia** will be formed and the equilibrium will be re-established.

Similarly, if extra ammonia is added, more nitrogen and hydrogen will be formed.

If the **pressure** is *increased* more ammonia will be formed. This is because forming more ammonia leads to a *lowering* of the pressure. In this way equilibrium is re-established.

A change of **temperature** leads to a change in the proportions of the equilibrium mixture. For **exothermic** reactions (e.g. Haber process) *raising* the temperature favours the *right → left* reaction (ammonia would react to form more nitrogen and hydrogen). This is because the right → left reaction is **endothermic**. It therefore counters the rise in temperature.

Lime This is **calcium oxide** (CaO). It is produced by heating **calcium carbonate**. It is used to make **slaked lime** and is responsible for turning impurities in the **blast furnace** into *slag*. In the laboratory it is a useful drying agent for **ammonia**.

Limestone This is a commonly found rock which contains between 50% and 90% **calcium carbonate** ($CaCO_3$). It is used to make **lime** and **cement** and is used as building stone and hard core for foundations.

Limewater This is a dilute solution of the sparingly soluble compound **calcium hydroxide**. It is an **alkali** and is used to test for **carbon dioxide**:

$$Ca(OH)_2(aq) + CO_2(g) \rightarrow CaCO_3(s) + H_2O(l)$$

A white precipitate is formed. If excess carbon dioxide is bubbled through, the precipitate reacts to form the soluble salt **calcium hydrogencarbonate**:

$$CaCO_3(s) + H_2O(l) + CO_2(g) \rightarrow Ca(HCO_3)_2(aq)$$

Linear molecules These are molecules which are *straight*, i.e. their **atoms** are *in a line*. All molecules which contain only two atoms must be linear but some examples of more complicated molecules are shown here:

carbon   ethyne   beryllium
dioxide   chloride
O=C=O   H—C≡C—H   Cl—Be—Cl

Liquid A **state of matter** in which particles are loosely bonded by inter-molecular forces. A liquid always takes up the shape of its container. The particles in the liquid are not fixed in a rigid framework (**lattice**).

See **evaporation** and **boiling**.

Lithium Lithium is a **group I metal**. It is the least

reactive **element** in the group but has vigorous reactions with acids. It has a quiet reaction with water. It is soft and tarnishes readily.

It is possible that the metal will be very important in the future if electricity can be generated by **nuclear fusion** reactions.

The metal ion gives a red **flame test**.

Litmus This is extracted from lichen. It is used as an **acid–base indicator**:

acidic solution	pH 7	alkaline solution
red	purple	blue

Litre A litre is exactly 1 **dm³** (1000 **cm³**). It is the **unit** of **volume** in everyday life, e.g. for fruit juice and paint, but the *scientific unit* is the cubic decimetre, **dm³**.

Lone pair of electrons These are pairs of *outer shell* **electrons** which are not used in bonds within the compound. They can, however, be used for forming bonds with other compounds.

Example:

Three bonding pairs of electrons

Here the lone pair is used to form the ammonium ion

145

M$_r$ This is the symbol for the **relative molecular mass** of a **compound** or **molecule**. It can be calculated by adding together the **relative atomic masses (A$_r$)** of each atom within the molecule.

Example:

M$_r$ carbon monoxide (CO) = 12 + 16 = 28
M$_r$ water (H$_2$O) = (2 × 1) + 16 = 18

Macromolecule This term means *large* molecule. It is usually used to describe **molecules** with an **M$_r$** value greater than one thousand, i.e. **polymers** such as the **carbohydrates** and the plastics, e.g. **poly(ethene)**, and also molecules such as **proteins** and nucleic acids.

Magnesium Magnesium is a shiny grey **group II metal**. It is quite reactive giving vigorous reactions towards **acids**. It burns in **air** and will react with steam to release **hydrogen**. The metal is obtained by the **electrolysis** of molten magnesium chloride. Much of this is extracted from sea water.

The chief use of magnesium is in the production of low density **alloys** for use in the aircraft industry.

Magnesium compounds

Magnesium hydroxide $Mg(OH)_2$	This is obtained by treating sea-water with calcium hydroxide. It is used in pharmaceutical products such as *Milk of Magnesia* as a treatment for excess acidity in the stomach.
Magnesium carbonate $MgCO_3$	This is found naturally as magnesite and as Dolomite ($MgCO_3.CaCO_3$) and is used for making heat-resistant (refractory) materials.
Magnesium oxide MgO	This important refractory compound is made by heating the carbonate.
Magnesium chloride $MgCl_2$	After extraction from sea-water the molten compound is electrolysed to produce magnesium metal
Magnesium sulphate $MgSO_4.7H_2O$	*Epsom Salts.* These crystals are used to treat constipation. They are also used as a fire-proofing agent.

Malleable A malleable substance can be beaten or hammered into different shapes. The larger the crystals in the substance, the more malleable it is. Materials which are malleable are usually also **ductile**. **Metals** and **alloys** have these properties.

Maltose $C_{12}H_{22}O_{11}$. This is a **disaccharide** molecule. It is found as a breakdown product of **starch**. Maltose is **glucose dimer**.

Manganese Manganese is a **transition metal**. It is found naturally as the oxide (MnO_2) and is extracted either by electrolysis of the sulphate or by a **thermit reaction**. The chief use of the metal is in **alloys**, e.g. **steels** and **bronzes**.

Manganese compounds Manganese forms compounds where the metal has valency 2, 3, 4, 6 and 7. The most important ones are 2, 4 and 7.

| Manganese(IV) oxide
MnO_2 | This has an important use in the dry **battery**. It is used as a **catalyst** and oxidizing agent. |
| Potassium manganate(VII)
$KMnO_4$ | This is a vivid purple crystalline substance which is a vigorous **oxidizing agent**. It is used as an antiseptic. It gives the MnO_4^- ion. |

Mass This is how much material a substance possesses. It is usually measured in **grammes** (g) or kilogrammes (kg).

Mass number The mass number of an **isotope** is the number of **protons** and **neutrons** in the **nucleus** of the atom. This number is shown at the top left-hand side of the symbol when describing the isotope.

Examples
$$_1^1H \quad _6^{12}C \quad _{11}^{23}Na \quad _{20}^{40}Ca \quad _{82}^{208}Pb \quad _{92}^{238}U$$

Melting A **solid** melts to form a **liquid** when the

lattice   lattice   (liquid)
(solid)   break down

energy of the particles is sufficient to break up the bonds holding them into a **lattice**. For a *pure* substance, melting occurs at a fixed temperature – the **melting point**.

Some bonds remain but clusters of particles are mobile.

Melting point The temperature at which solid and liquid forms of the same substance (e.g. ice/water) are in equilibrium. At constant pressure, the melting point is a constant for a **pure** substance but it is *lowered* if impurities are added.

Mercury Mercury is the only liquid **metal** at room temperature. It is used to make **amalgams** and in **electrolysis** cells as a **cathode**. Its compounds are very poisonous.

Metal 17% of elements are metals. A list of their general properties is shown below. Pure metals are rarely used today. Most metals are used in the form of **alloys**. These play a vital part in our lives.

Metals produce **cations**, react with **acids**, are strong and hard, are **malleable** and **ductile** and react with non-metals. They have a crystalline structure, are shiny and have high **melting** and **boiling points**.

Metalloid These elements are neither **metals** nor **non-metals**. They are sometimes referred to as *semi-metals*.

The best examples are germanium and **arsenic**. Such substances tend to have the physical properties of metals, e.g. have a shiny appearance and a high melting and boiling point but the chemical properties of non-metals, e.g. do not react with acids. They are useful as **semi-conductors**.

Methane Methane is a gaseous **alkane**. It is the main constituent of **natural gas**. It is also released from **petroleum** whilst it is being processed. It burns readily to give carbon dioxide and water and is an industrial source of **hydrogen**.

CH₄

Methanol Methanol is an **alcohol**. It is a colour-

CH₃OH

less poisonous liquid. It is added to **ethanol** to make **methylated spirits**.

Methylated spirits This is a mixture of **ethanol** (90%), **methanol** (9.5%), pyridine (0.5%) and traces of a purple dye. This mixing is done to make the ethanol unfit for drinking. There is no excise tax on the mixture and so it is inexpensive. It is used as a fuel and a **solvent**.

Methylbenzene This is an **aromatic hydrocarbon** which is produced from petroleum. It is used to make TNT the explosive (TriNitroToluene) and is a useful **solvent**. Its old name is toluene.

Methyl orange This is an **acid-base indicator**.

| acidic | alkaline |
| solution | solution |

←   →

orange/red   yellow

Miscible liquids These are liquids which can mix together completely. They can dissolve in each other. See **Immiscible**.

Mixture When two or more substances are present in the same container they are said to be a mixture, e.g. **baking powder, air**, gunpowder, **petrol, methylated spirits**.

Properties of mixtures

Mixtures can be made in all proportions.

Making mixtures does not involve the release or absorption of heat.

The properties of a mixture are the properties of *all* the components.

Mixtures can be separated by physical means.

mmHg This symbol represents the **unit** *millimetres of mercury*. It is a unit of **pressure**. One **atmosphere** pressure equals 760 mmHg. This is the **air** pressure which would support a column of mercury 760 mm high.

Molarity The molarity of a **solution** is its **concentration** expressed in **moles** per cubic decimetre of solution (**mol/dm^3**) See **Molar solution**.

Molar solution A molar solution is one where the **concentration** of **solute** is *one* **mole** per cubic decimetre of *solution* 1 **mol/dm^3**, i.e. a solution where one mole of solute has been dissolved in the **solvent** and then sufficient solvent has been added to make 1000 **cm^3** of solution.

Mole This is the amount of a substance (**element** or **compound**) which contains **L** particles (L is the **avogardro constant**).

L is defined as the number of atoms there are in 12 g of the **carbon**-12 **isotope**. It follows that the value of a mole for individual atoms and compounds is the **relative atomic mass** (expressed in grams).

Examples:
One mole of

Oxygen	(O_2)	is $16 + 16$	$= 32$ g
Water	(H_2O)	is $2(1) + 16$	$= 18$ g
Ammonia	(NH_3)	is $14 + 3(1)$	$= 17$ g
Ethene	(C_2H_4)	is $2(12) + 4(1)$	$= 28$ g

mol/dm^3 This is a **unit** of **concentration**. 1 mol/dm^3 means that one **mole** of substance would be present in one cubic decimetre (**dm^3**) of the **solution**. The unit is sometimes written mol dm^{-3}.

Molecule A molecule is defined as the smallest particle of an **element** or **compound** which exists independently. It contains **atoms** bonded together in a fixed whole number ratio.

Oxygen O_2	Phosphorus P_4
Nitrogen N_2	Carbon dioxide CO_2
Neon Ne	Water H_2O
Sulphur S_8	Hydrogen chloride HCl
Hydrogen H_2	

All organic compounds are molecules.

154

Molecular formula The molecular formula of a substance shows the number and types of atoms in the molecule. It tells us nothing about how the atoms are arranged, e.g. $C_4H_{10}O$ is the molecular formula of butanol and **ether** (ethoxyethane). See **Empirical formula**.

Monatomic molecule A **molecule** which only contains one **atom** is termed monatomic. Only the **inert gases** are monatomic. Because of their **electronic configurations** it is difficult for inert gas atoms to form bonds, and so they exist in the monatomic form.

Monobasic acid A monobasic acid only contains *one* **hydrogen** atom per molecule which can be replaced by a metal. Only normal salts can be formed – not **acid salts**

Examples include:
   Hydrochloric acid HCl   **Ethanoic acid**
   Nitric acid HNO_3

Monomer A monomer is the compound from which **polymers** are made, e.g. **ethene** $CH_2 = CH_2$ forms **poly(ethene)** $(CH_2 - CH_2)_n$.

Monosaccharide This is the simplest kind of **sugar**. In these compounds there is a *single* molecule

155

as opposed to two (or more) which have reacted together to form **disaccharides** or **polysaccharides**.

Glucose	$C_6H_{12}O_6$
Fructose	$C_6H_{12}O_6$
Ribose	$C_5H_{10}O_5$

Multiple bond Bonds which contain two **electrons** are called single bonds. Those which contain four or six electrons are multiple bonds. **Double bonds** contain four electrons. **Triple bonds** contain six.

Examples of compounds which contain multiple bonds:

Ethene C_2H_4 – **double bond**
Ethyne C_2H_2 – **triple bond**

Nanometre This is a unit of length. It is 10^{-9} of a metre, i.e. 100000000 make one metre. It is useful in measuring the length of bonds, e.g. H–H bond length = 0.074 nm.

Naphtha This is a mixture of **hydrocarbons** which is produced by the **fractional distillation** of **petroleum**. The **boiling points** of the compounds in the mixture are in the range 80–160°C. Naphtha is usually subjected to **cracking. Ethene** is an important product of this process. Naphtha can be oxidized to give **ethanoic acid**.

Natural gas This is a mixture of mainly **hydrocarbon** gases which is found in deposits beneath the earth's surface. Natural gas and petroleum are often found together. **Methane** is usually the major constituent. The **inert gas, helium**, is sometimes present in the mixture.

It is used as a fuel both in industry and the home.

Neon Neon is an **inert gas** which is found in the **atmosphere** to a small extent. It forms no known compounds but is used to fill fluorescent tubes. A red glow is produced which is used for advertising signs.

Neutral A neutral **solution** is one which is neither acidic nor alkaline. Neutral solutions contain the same concentration of **hydroxide** and **oxonium ions**. The pH of a neutral solution is seven at 25 °C.

A neutral **oxide** does not react with either **acids** or **alkalis**. The best example is **water**.

A neutral particle is one which does not have an electric charge, e.g. a **neutron**.

Neutralization This is the process in which either the **pH** of an acidic solution is increased to seven or the pH of an alkaline solution is decreased to seven. The resulting solution contains the same concentration of **oxonium** and **hydroxide** ions i.e. it is neutral.

An acid can be neutralized by the addition of a base or a compound such as a carbonate.

$$HCl(aq) + NaOH(aq) \rightarrow NaCl(aq) + H_2O(l)$$
$$2HCL(aq) + CaCO_3(s) \rightarrow$$
$$CaCl_2(aq) + CO_2(g) + H_2O(l)$$

Acid/base **indicators** show when neutralization is complete.

Neutron This is one of the particles which are found in the **nucleus** of all atoms except hydrogen. It has approximately the same mass as the **proton** but *no* charge. See **Nuclear reactions**.

Nickel Nickel is a transition metal. It is a magnetic substance which is found in nature as the **sulphide** and is oxidized to the **oxide**, reduced by

58.7

28   Ni   8.9

1455   2837

2.8.16.2

hydrogen and then purified by using carbon monoxide gas:

$$Ni(s) + 4CO(g) \rightarrow Ni(CO)_4(g)$$

This reaction can be reversed at high temperatures, producing **pure** nickel.

Nickel can have a **valency** of 2, 3 or 4, but only nickel(II) salts are common. Nickel is used as a **catalyst**, in **alloys** (nichrome, coinage metal, stainless steel) and in plating.

Nitrate Nitrates contain the $-NO_3$ group with **valency** = 1. These are **salts** of **nitric acid** and can be made by treating a metal **carbonate** or **oxide** with the dilute acid:

$$CuCO_3(s) + 2HNO_3(aq) \rightarrow$$
$$Cu(NO_3)_2(aq) + CO_2(g) + H_2O(l)$$
$$ZnO(s) + 2HNO_3(aq) \rightarrow$$
$$Zn(NO_3)_2(aq) + H_2O(l)$$

Nitrates are easily decomposed by heat. There are three kinds of reaction:

1) $2KNO_3(s) \rightarrow 2KNO_2(s) + O_2(g)$
examples:
Sodium, Potassium.

2) $2Cu(NO_3)_2(s) \rightarrow 2CuO(s) + O_2(g) + 4NO_2(g)$
all metal nitrates other than shown in 1 and 3.

3) $2AgNO_3(s) \rightarrow 2Ag(s) + 2NO_2(g) + O_2(g)$
examples: *Silver, Mercury.*

Nitrates are important fertilizers, e.g. **sodium nitrate** and **ammonium nitrate**. If used incorrectly they can lead to serious water pollution.

Nitric acid (HNO_3) This colourless, corrosive liquid is made from **ammonia** by **oxidation** over a **platinum**/rhodium **catalyst**. There are three stages:

$$4NH_3(g) + 5O_2(g) \rightarrow 4NO(g) + 6H_2O(g)$$
$$4NO(g) + 2O_2(g) \rightarrow 4NO_2(g)$$
$$4NO_2(g) + 2H_2O(l) + O_2(g) \rightarrow 4HNO_3(aq)$$

The acid is distilled to a concentration of 68%

The acid can be prepared in the laboratory by heating a nitrate with *concentrated* **sulphuric acid**:

$$KNO_3(s) + H_2SO_4(l) \rightarrow KHSO_4(s) + HNO_3(g)$$

Nitric acid is a vigorous oxidizing agent used in the production of **fertilizers** and explosives. The **salts** of nitric acid are termed **nitrates**. See **Nitrogen oxides** and **Nitrites**.

Nitrites These are salts of nitrous acid (HNO_2). The sodium and potassium salts can be made by heating the nitrate:

$$2NaNO_3(s) \rightarrow 2NaNO_2(s) + O_2(g)$$

They are used in the preservation of meats.

Nitrogen (N_2) Nitrogen is a non-metallic element in group v of the periodic table. It is an unreactive, **diatomic** gas which forms about 78%

of the **atmosphere**. It is produced by the **fractional distillation** of liquid **air**.

Nitrogen is an important chemical because of the need for **nitric acid** and **ammonia**. It is *fixed* from the atmosphere in the **Haber process**.

See **Nitrogen oxides** and **Nitrogen cycle**.

Nitrogen cycle

The diagram shows the ways in which animals, plants and Man are involved in the chemical links between **nitrogen**, **ammonia** and **nitrates**. Nitrogen is *fixed* from the **atmosphere** by:

1) The **Haber process**.
2) Electrical discharges (lightning).
3) Action by soil bacteria.

The nitrates in the soil are taken up by plants which are then eaten by animals. Both animal and plant **protein** can rot and revert to ammonia. Animals also excrete urine which decomposes to ammonia.

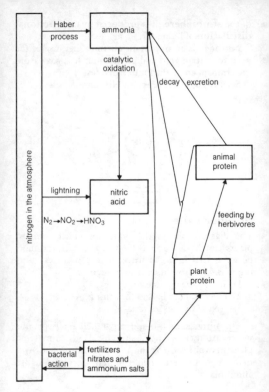

162

Nitrogen is returned to the atmosphere by the action of other bacteria.

Despite Man's extraction of **nitrogen** from the air in the Haber process the amount of nitrogen in the atmosphere remains approximately constant.

Nitrogen oxides

Nitrogen monoxide NO	This is a colourless gas which can be produced by the action of *moderately* concentrated **nitric acid** on **copper**. The gas reacts *immediately* with oxygen to form nitrogen dioxide: $$2NO(g) + O_2(g) \rightarrow 2NO_2(g)$$
Nitrogen dioxide NO_2	This is a brown gas. It has a choking smell and irritates the lungs and windpipe. Breathing it can lead to death from pneumonia. It can be produced in the laboratory by the action of concentrated **nitric acid** on **copper**.
Nitrous oxide N_2O	This is a colourless gas with a sweetish smell. It is an anaesthetic and is used in dentistry. It is sometimes called laughing gas.

Noble gases See **Inert gases**.

Non-metal   Non-metals are elements which have (a) either molecular structures, and thus are **gases** at **room temperature**, or are **solids** or **liquids** with low **melting** and **boiling points**; or (b) **giant structures** with **covalent bonding**. Typical properties of non-metals are:

Poor **conductivity**.

Do not react with **acids**.

Produce **acidic oxides**.

Form **covalent** compounds.

Give rise to **anions**.

Examples:
The halogens, the inert gases, oxygen, sulphur, carbon, nitrogen.

N.T.P.   See **S.T.P.**

Nuclear reactions   Nuclear reactions are very different from normal chemical reactions. New **compounds** are not formed but changes occur in the nuclei and new *elements* can be formed. Most elements have stable and unstable isotopes, e.g.

carbon-12 carbon-13   carbon-14

stable   unstable

But some have no stable isotopes, e.g. **uranium, plutonium**. Whether an isotope is unstable depends on the numbers of **neutrons** and **protons** in the **nucleus**. When an isotope is unstable, several reactions can occur:

1) The isotope can *split* – fission can occur. This creates two stable isotopes. Neutrons are also released and lots of energy is released.

2) A neutron can decay into a proton and an **electron**:

$$_0^1n \rightarrow {}_1^1p + {}_{-1}^0e$$

The electron is expelled from the nucleus (**beta particle**) and the **atomic number** of the atom *increases* by 1, e.g:

$$_{82}^{209}Pb \rightarrow {}_{83}^{209}Bi + {}_{-1}^0e$$

There is no change in mass.

3) An **alpha particle** is expelled from the atom, e.g:

$$_{92}^{238}U \rightarrow {}_{90}^{234}Th + {}_2^4He$$

4) An unstable atom can emit **gamma rays**. This often occurs after the emission of a **beta particle**. See **Half-life**.

These nuclear reactions have been used by Man in many ways. Nuclear power stations use the heat produced in the fission of uranium, plutonium or thorium isotopes. The heat raises steam which drives a turbine producing electricity. This same source of energy is also used by some countries in the form of bombs. These have terrible destructive powers.

Isotopes producing gamma rays are used to destroy bacteria in food processing and *cancerous* cells in the body (chemotherapy).

Isotopes are also used in industry for a variety of

165

analytical purposes, e.g. detecting cracks in pipelines.

Man is striving to produce useful energy from **fusion** reactions and if this is possible it will provide a source of **electricity** which is not dependent upon **coal** or **petroleum**.

The products of radioactive decay (**radioactivity**) are dangerous as they can produce harmful effects on the body including *leukaemia* and *cancers*. Great care has to be taken when using radioactive materials.

Nucleus This is the part of an **atom** where the **mass** is concentrated. It contains **protons** and **neutrons** and is usually pictured as being the compact centre of a spherical atom. **Electrons** move *round*

a nucleus of
carbon – 12 contains
6 protons and 6
neutrons

isotopes have
different nuclei.

a nucleus of
carbon – 14 contains
6 protons and 8
neutrons

the nucleus. The nucleus has a positive charge and in the neutral atom this is balanced by the charges on the electrons.

Hydrogen-1 **isotopes** are the only atoms whose nuclei contain *no* neutrons. See **Nuclear reactions**.

Nylon The nylons are a family of **polyamide polymers**. The most common of them is *Nylon 6.6*. This *man-made fibre* has great strength. Its uses include fabrics (shirts, cloth), yarns (stockings knitwear), carpets, ropes and nets. It is useful because it will not rot, it does not absorb water but it does stretch. This is useful in ropes and stockings. It is often mixed with other fibres e.g. wool, to get the correct balance of properties.

6 carbon atoms   6 carbon atoms

Nylon 6.6

Octane rating **Petrol** and **air** mixtures have to explode at exactly the correct moment in an internal combustion engine. If the wrong kind of fuel is used ignition of the mixture can occur *before* it should. This leads to the characteristic sound of *pinking* (or *knocking*).

The octane rating (or octane *number*) of a petrol is

a measure of the *anti-knock* qualities of the fuel. The higher the rating, the better it is. Four star petrol has an octane rating of ninety-eight. The greater the proportion of *branched* **hydrocarbon** molecules in the fuel, the greater the rating will be.

Octane ratings can be raised by adding substances such as tetraethyl lead. This unfortunately leads to lead pollution in the atmosphere.

Oil   An oil is a liquid fat, e.g. melted butter, olive oil, sunflower oil, etc.

The word is sometimes used in place of **petroleum**.

Oleum   This is a solution of **sulphur(VI) oxide** in *concentrated* **sulphuric acid**. It is a very corrosive substance and is a vigorous **oxidizing agent**. See **Contact process**.

Ore   An ore is a naturally-occurring substance from which an **element** can be extracted, e.g:

galena – lead, haematite – iron, blende – zinc

Organic chemistry   This is the study of **compounds** of **carbon**. It does not include carbonates, carbon dioxide, etc. See **Inorganic chemistry**.

Oxidation   A substance undergoes oxidation if it:

gains oxygen: $2Mg(s) + O_2(g) \rightarrow 2MgO(s)$
loses hydrogen: $CH_4(g) + Cl_2(g) \rightarrow$
                         $CH_3Cl(l) + HCl(g)$
loses electrons: $Cu(s) \rightarrow Cu^{2+}(aq) + 2e^-$

See **Oxidizing agent**, **Reduction** and **Redox**.

Oxide   An oxide is a compound between an **element** and **oxygen** only. Oxides can be formed by:
   a) direct combination of the elements
   b) oxidizing compounds
   c) the action of heat on a **carbonate**, an **hydroxide** or some **nitrates**.

Examples:
$$2Mg(s) + O_2(g) \rightarrow 2MgO(s)$$
$$CH_4(g) + 2O_2(g) \rightarrow CO_2(g) + 2H_2O(g)$$
$$CuCO_3(s) \rightarrow CuO(s) + CO_2(g)$$
$$Pb(OH)_2(s) \rightarrow PbO(s) + H_2O(g)$$
$$2Zn(NO_3)_2(s) \rightarrow 2ZnO(s) + 4NO_2(g) + O_2(g)$$

Oxides can be reduced to the element with a suitable **reducing agent**, e.g. hydrogen, carbon or carbon monoxide. See **Basic oxide, Acidic oxide, Amphoteric oxide**.

Oxidizing agent   Something which causes the **oxidation** of another substance. When this happens the oxidizing agent is reduced.

Common oxidizing agents:

Oxygen	O_2
Chlorine	Cl_2
Ozone	O_3
Hydrogen peroxide	H_2O_2
Potassium manganate(VII)	$KMnO_4$

Oxidizing agents at work:

$$Mg(s) + O_2(g) \rightarrow 2MgO(s)$$
$$Cl_2(g) + 2Na(s) \rightarrow 2NaCl(s)$$
$$H_2O_2(aq) + H_2S(g) \rightarrow 2H_2O(l) + S(s)$$

Oxonium ion (H_3O^+) The loss of an **electron** from a **hydrogen atom** leads to the formation of a hydrogen ion. This is a proton.

The proton is such a small, reactive particle that in **aqueous** solution it forms bonds to water molecules. More than one water molecule is probably involved but, for simplicity only one is usually shown.

$$H \rightarrow H^+ + e^-$$

$$H^+ \cdots\cdots O \overset{\displaystyle H}{\underset{\displaystyle H}{}}$$

$$H_3O^+(aq)$$

It can be used in ionic equations in place of $H^+(aq)$:

$$H_3O^+(aq) + OH^-(aq) \rightarrow 2H_2O(l)$$
$$\text{acid} \quad + \quad \text{alkali}$$

Oxygen (O_2) Oxygen is a gaseous, non-metallic **element** in **group** VI of the **periodic table**. It makes up 21% of the **atmosphere**. It is a vigorous **oxidizing agent** and is vital for the **respiration** of plants and animals.

It is a reactive gas, readily forming **oxides** with

most **elements**. It is obtained industrially by the liquifying of air and in the laboratory by the decomposition of **hydrogen peroxide**:

$$2H_2O_2(aq) \rightarrow 2H_2O(l) + O_2(g)$$

A **catalyst** such as manganese(IV) oxide is usually used.

Oxygen is a colourless, odourless **diatomic** gas. It is neutral and is only slightly soluble in water (40 cm³ per cubic decimetre). It is sufficiently soluble to allow fish and other aquatic life to live in water. Problems arise if the oxygen in water drops to too low a level. Then, fish suffocate and die. Such

splint bursts into flames

a drop in the **concentration** of oxygen is usually caused by pollution.

The chemical test for oxygen is to plunge a glowing splint into the gas. If the gas is oxygen, the splint will relight. The gas will also make anything which is already burning in air, burn much more fiercely.

Ozone (O_3) This is an **allotrope** of **oxygen** where the molecule is *tri*atomic $O_3(g)$. It is a blue gas and is a very powerful **oxidizing agent**. It occupies a very small part of the **atmosphere** and is mainly located about 30 km above the surface – the *ozone layer*. Here it prevents most of the **ultraviolet radiation** from reaching the earth where it could do great damage. It can be formed from oxygen by passing a spark through the gas:

$$3O_2(g) \rightarrow 2O_3(g)$$

Paraffin This is a fuel which is obtained from **petroleum**. It has a boiling range of 160–250°C. See **Kerosine**.

The term was once used instead of **alkane**.

Parafin wax consists of a mixture of solid **hydrocarbons**. It is used to make candles.

Pascal This is the **SI unit** of **pressure**. One **atmosphere** pressure is about one hundred kilopascals, (100 kPa). One pascal is equivalent to a force of one Newton on one square metre (1 Nm^{-2}).

Passive Describes a metal which possesses a surface layer of **oxide** which makes it unreactive – passive. **Aluminium** has such a layer and **iron** can be given one by dipping the metal into concentrated **nitric acid**.

Peptide When two or more **amino acids** react together, peptides are formed. If three or more amino acids are involved the term **polypeptide** is usually used. **Proteins** consist of long polypeptide chains which are often linked together in a variety of ways.

Peptides contain the *peptide link*:

Percentage composition In analysis it is often useful to know what proportion of a compound is made up of any given element. This is usually expressed in percentage terms:

Examples	% by *mass* of	
Methane	carbon	hydrogen
CH_4	75%	25%
Ethyne C_2H_2	92%	8%

Knowing the percentage composition of a compound it is possible to work out its **empirical formula**.

Example:
A compound is known to have a percentage composition of: Ca 40% C 12% O 48%.

To find the *molar* ratio of the elements in the compound, divide each figure by the respective A_r value (Ca) $^{40}/_{40} = 1$ (C) $^{12}/_{12} = 1$ (O) $^{48}/_{16} = 3$.

The molar ratio is: $1 : 1 : 3$

The empirical formula is: $CaCO_3$

Period In the **periodic table** the horizontal rows of **elements** are known as periods. There are seven in all. See **group**.

Periodic table The periodic table which is shown below is a way of presenting all the **elements** so as to show their similarities and differences.

The elements are arranged in increasing order of **atomic number (Z)** as you go from left to right across the table.

Period 3

Na	Mg	Al	Si	P	S	Cl	Ar
11	12	13	14	15	16	17	18

The horizontal rows are called **periods** and the vertical rows, **groups**. An **inert gas** is found at the right hand side of each period.

Period	1	He
	2	Ne
	3	Ar
	4	Kr
	5	Xe
	6	Rn

There is a progression from metals to non-metals across each period.

Similar elements are found in a group, e.g. these elements have a similar **electronic configuration**.

group I	alkali metals
	Li Na K

group VII	halogens
	F Cl Br I

The number of **electrons** in the outer shell is the same as the number of the group, e.g. group I.

lithium	2·1
sodium	2·8·1
potassium	2·8·8·1

Elements 58 to 71 are known as the *Lanthanide* or *Rare Earth* elements. These elements are found on earth in only very small amounts.

176

Elements 90 to 103 are known as the *Actinide* elements. They include most of the well known elements which are found in **nuclear reactions**. The elements with larger atomic numbers than 92 do not occur naturally. They have all been produced by man by bombarding other elements with particles. **Plutonium** is formed in nuclear reactors.

Elements with atomic numbers of 104 and above are not being named after famous scientists or places. They are being named in a systematic way according to their atomic number:

element 104 is *Unnilquadium*   Un = 1 Symbol
                                    nil = 0   *Unq*
                                    quad = 4

element 118 would be *Ununoctium* un = 1
                                    un = 1   *Uuo*
                                    oct = 8

Peroxide A peroxide contains the O_2^{2-} **ion** or $-O_2$ group. The best example is **hydrogen peroxide** (H_2O_2). The peroxides of alkali metals are also known. Hydrogen peroxide can be made by the action of acid on these:

$$Na_2O_2(s) + 2HCl(aq) \rightarrow 2NaCl(aq) + H_2O_2(aq)$$
or $O_2^{2-}(s) + 2H^+(aq) \rightarrow H_2O_2(aq)$

Peroxides are vigorous **oxidizing agents**.

Petrochemical This is a chemical which has been made from petroleum. Examples would include **ethene** and propene. These chemicals are used in the manufacture of other chemicals, e.g. **poly(ethene)** and **poly(propene)**. Petrochemicals are *intermediates* in the production of finished products.

Petrol This is a mixture of **hydrocarbons** which is used as a fuel in internal combustion engines. It is produced from **petroleum** in a refinery. See **octane rating**. Petrol is also known as gasoline or motor spirit.

Petroleum This is the term describing the mixture of **hydrocarbons** which are found in the earth's crust, e.g. **natural gas** and *crude oil*. It has been produced over millions of years from the remains of *marine* animals and plant organisms. It is the raw material for the petrochemical industry and is the source of our **petrol, diesel** fuel, heating oil, fuel oil and gas supplies. Vast reserves of petroleum are found in the Middle East, the United States, the Soviet Union, Central America and the North Sea.

Petroleum has replaced **coal** as the chief source of raw materials for the chemical industry. But the supply of petroleum will not last forever and the search is now on for the substance which will replace it in our lives.

How petroleum is converted into useful products is described in **refining**. See also **Fractional distillation**.

Pewter This is an **alloy** which was made of **lead** and **tin** (80–90%) with small amounts of antimony. It was widely used for plates and mugs and is now used as jewellery. The pewter made now contains no lead because of the danger of the lead getting into solution or contaminating food. It has been replaced by **copper** and antimony.

pH

acidity increasing   neutral   alkalinity increasing

← ——————————————————————— →

| 0 | 3 | 7 | 10 | 14 |

The pH scale is a measure of the acidity or *alkalinity* of a solution. The lower the value, the more acidic is the solution, i.e. the larger the **concentration** of **oxonium ions** there are within it. A **neutral** solution, where the concentrations of oxonium and hydroxide ions are equal, has a pH of 7 at 25 °C.

$$pH = -\log_{10}[H_3O^+]$$

The pH of a solution is the negative logarithm (base 10) of the concentration of oxonium ions (mol/dm³)

Example:

The pH of a solution whose concentration is 0.1 mol/dm³ is $-\log(0.1) = 1$

pHs of common substances	
substance at a concentration of 1 mol/dm³	pH
strong acid (HCl)	0
weak acid (CH$_3$COOH)	4
water	7
ammonia solution	10
strong alkali (NaOH)	14

Phase change (change of state)   A phase change occurs when a substance goes from one physical state to another. The possible phases in a chemical system are shown here:

a and c are condensations
b is freezing
d is vaporization
e is melting
f is sublimation

Phenol

$$C_6H_5OH$$

Phenol is a white crystalline **aromatic** solid.

It is a very corrosive chemical and is poisonous. It is useful as an antiseptic, and **disinfectant** and many household products are based on similar molecules to phenol, i.e. compounds with similar (or better) disinfectant properties but which are less corrosive, e.g. T.C.P. and Detol.

The O – H group in phenol is *acidic* and salts can be formed:

$$C_6H_5O^-Na^+(s)$$

Phenolphthalein This is an **indicator** which is used to follow **acid–base** reactions. The molecular structure of the compound is based on the **phenol** molecule.

acidic solution	alkaline solution
colourless	red

← →

181

Phosphates These are **salts** of **phosphoric (v) acid**. Phosphoric acid is a *tri*basic acid and, therefore, gives rise to three kinds of salts. These are shown here:

sodium phosphate	Na_3PO_4
sodium hydrogenphosphate	Na_2HPO_4
sodium dihydrogenphosphate	NaH_2PO_4

Phosphates are used in fertilizers to replace the **phosphorus** – containing compounds in the soil. It comes in several forms:

Superphosphate – a mixture of calcium sulphate and calcium dihydrogenphosphate ($Ca(H_2PO_4)_2$).

A mixture of ammonium nitrate and ammonium hydrogenphosphate ($(NH_4)_2HPO_4$ is made by reacting **ammonia** with a mixture of **phosphoric (v)** and **nitric acids**.

Calcium phosphate is the chief constituent of animal bones and phosphates are used extensively in washing powders and detergents.

Slag from **blast furnaces** is put directly onto the soil.

Phosphoric(v) acid (H_3PO_4) This is a tribasic acid which gives rise to three kinds of **phosphate**. It is a **solid** at **room temperature** but is usually sold as a viscous **solution** in **water**.

Phosphorus (P_4) Phosphorus is a solid non-metallic **element** in group v of the **periodic table**. Three **allotropes** exist: white, red and black phosphorus. The information in the chart refers to *white* phosphorus.

The allotropes has very different physical properties, e.g. **crystal** structure, **density, melting** and **boiling points**.

White phosphorus is very poisonous and has to be kept under water because it bursts into flame when it comes into contact with air, forming the oxide:

$$P_4(s) + 5O_2(g) \rightarrow P_4O_{10}(s)$$

Rocks containing phosphorus compounds are quite common in the earth's crust and phosphorus is an important element for the maintenance of life. Plants need it and so it is a necessary component of fertilizers. Some enzymes used in **respiration** also contain the element. Animal bones contain **phosphates**.

Photochemical reactions These are chemical reactions which are brought about by light. The process of photography depends on light. When silver bromide is exposed to light it decomposes:

$$2AgBr(s) \rightarrow 2Ag(s) + Br_2(g)$$

The black part of the negative of a film is a thin layer of **silver**. This is then processed to produce the final pictures.

Photosynthesis This is the most important photochemical reaction. Our existence on this planet depends on it.

Other examples are the reaction between **hydrogen** and **chlorine** which can be started off by a bright light e.g. burning magnesium:

$$H_2(g) + Cl_2(g) \rightarrow 2HCl(g)$$

In hot areas of the world, e.g. California, where there are a lot of vehicle exhaust fumes, photochemical *smog* can be produced by the reaction of the exhaust fumes. This is an acrid haze which is very unpleasant.

The **bleaching** action of the sun is also a photochemical reaction as is the conversion of cholesterol in the skin into **vitamin** D.

In all these reactions light provides the **energy** that makes the reactions work.

Photosynthesis This is a **photochemical** process in which **carbon dioxide** and **water** are converted into **carbohydrate** and **oxygen**:

$$6CO_2(g) + 6H_2O(l) \rightarrow C_6H_{12}O_6(s) + 6O_2(g)$$

This occurs in the leaves of plants and is the means whereby plants obtain their *food*. It also puts **oxygen** back into the **atmosphere**.

The reaction is catalysed by **chlorophyll** and sunlight provides the **energy** for the reaction.

Physical change A physical change to a substance involves changes in its physical **properties** with no alterations to chemical properties.

Physical chemistry This is the study of the physical **properties** of elements and compounds and the relationship between their chemical properties and physical properties.

Pipette

A pipette is a piece of glassware used for measuring and transferring a fixed **volume** of liquid, e.g. 25 cm^3.

The liquid is drawn up into the pipette through the jet by suction applied at s. Liquid is taken up until the bottom of the meniscus is on the graduation.

To remove the liquid from the pipette, the suction is released and the jet is touched against the side of the vessel into which the liquid flows. In these circumstances, the pipette will release exactly the correct volume of liquid.

A safety pipette filler should always be used to fill a pipette. See **Burette**.

Planar A planar molecule is one which has all its atoms in the same plane. All **linear** and *tri*atomic molecules must be planar. Boron trifluoride is planar

$$F\!-\!\underset{\underset{F}{|}}{B}\!-\!F$$

Planar molecules

H_2O	C_2H_2	CO	HCl
CO_2	SiO_2	NO_2	N_2

Most carbon compounds are *non-planar* because of the **tetrahedral** arrangement of atoms around the carbon atom. Most alkenes are non-planar but **ethene** is an exception.

Propene is *non-planar* because the three hydrogen atoms attached to the carbon atom are above or below the other atoms.

ethene is planar   but propene is not

Plastic Capable of being shaped or moulded by **heat** and **pressure**. The word is usually used to describe man-made **polymers**, e.g. **poly(ethene)**, **nylon**, etc. Over thirty different types of polymers exist and by combining them it is now possible to produce a man-made substance with specific properties. See **Thermosetting**, **Thermoplastic**.

Platinum Platinum is a **transition metal** which is used as a **catalyst** in the chemical industry. It is also used as a metal for jewellery. It is expensive because it is useful and rare. Platinum occurs in nature as the element.

Plutonium   Plutonium is a man-made element. It is made in nuclear reactors from the 238 **isotope** of **uranium** by the absorption of a **neutron**:

$$^{238}_{92}U + ^{1}_{0}n \rightarrow ^{239}_{92}U \text{ this is unstable}$$

$$^{239}_{92}U \rightarrow ^{239}_{93}Np + ^{0}_{-1}e \text{ (\textbf{beta particle})}$$

Neptunium–239 is also unstable:

$$^{239}_{93}Np \rightarrow ^{239}_{94}Pu + ^{0}_{-1}e \text{ (beta particle)}$$

Uranium–239 and neptunium–239 have **half-lives** of twenty minutes and two days respectively, and so soon decay. The half-life of plutonium–239 is over 24 000 years and so is fairly stable.

Plutonium is used in bombs and as a fuel for nuclear reactors, e.g. *fast breeder reactors*. The cores of these reactors are surrounded by uranium–238 which is connected to plutonium–239 (in the way shown above) while the reactor is working. In this way it is possible to produce more fuel than is used in the reactor.

Plutonium is a toxic and dangerous chemical and must be handled with great care. Great controversies surround its use.

Poison A poison is a substance which is harmful if taken into the body and which will lead to death or injury. Poisons can be breathed in, e.g. hydrogen cyanide (HCN), **carbon monoxide** (CO) or lead fumes in the air; they may be eaten or drunk, e.g. arsenic(III) oxide, strychnine, or paraquat or they may pass into the body through the skin, e.g. **benzene** and nerve gases.

Radiation from nuclear **isotopes** can also be viewed as a poison.

Polar bond This is a **covalent bond** in which the **electrons** are not equally shared. This results in slight charges on the atoms, e.g.

$$\delta+ \quad \delta-$$
$$H - Cl$$

In hydrogen chloride, there is a small *positive* charge on the hydrogen atom and a small **negative** charge on the chlorine atom because of the greater attraction of electrons to the chlorine. Compounds with polar bonds tend to be soluble in compounds with polar bonds, e.g. water.

Polyamides These are man-made **polymers** which contain the *amide* group of atoms. Some compounds are usually known by the name **nylon**. Polyamides are *condensation* polymers.

amide group

Poly(chloroethene) (PVC) This **polymer** is made from chloroethene ($CH_2 = CHCl$). PVC has a wide range of uses: vinyl flooring, food containers, records, fibres, guttering.

Polyesters These are man-made **polymers** which contain an **ester** group of atoms. A well-known polyester is **Terylene** which is widely used in everyday life. They are *condensation* polymers whose **monomers** are an **acid** and an **alcohol**.

ester
group

Poly(ethene) This **polymer** is made from **ethene** and is ususally known as polythene.

$$(C_2H_4)n$$

Poly(ethene) is an **addition polymer**. Here, n is a very large number (about 50 000).

Poly(ethene) is widely used today. The polymer is a **saturated alkane** and so is very unreactive. It is used for containers, e.g. ice-cream or margarine tubs, for packaging, e.g. cling-films and plastic bags, and many other uses.

The **monomer**, ethene, is produced by **cracking** the products of **petroleum** refining, e.g. **naphtha**.

Polymers Polymers are large molecules in which a group of atoms are repeated, e.g.

$$x - x - x - x - x - x - x$$

or

$$x - y - x - y - x - y - x - y$$

191

Polymers are either naturally occurring or man-made. Starch and cellulose are natural ones whilst **nylon, Terylene, poly(chloroethene)** and **poly(propene)** are man-made.

Polymers are made by reacting **monomer** molecules together usually with a **catalyst**.

Addition polymerization involves one kind of unsaturated molecule. In the **polymerization** a long chain of atoms (linked by **single bonds**) is formed. See **Poly(ethene)**. Because the polymer is a **saturated** molecule it is usually fairly *un*reactive and this gives the polymer useful properties. See the entries for individual polymers.

Example

monomer (acrylonitrile)

polymer (used to make fabrics such as *Acrilan* and *Courtelle*)

Condensation polymerization involves two molecules which condense together into long chains. A small molecule is eliminated during the reaction. **Nylon** and **Terylene** are good examples of condensation polymers. Nylon is made by reacting an acid and an amine

The source of most of the raw materials used in making man-made polymers is **petroleum**. There has been great growth in the industry in recent years and will no doubt continue as long as petroleum or a substitute is available.

Polymerization This is the reaction of **monomers** to form a **polymer**.

Polypeptide A polypeptide is a molecule which is made up of up to fifty amino acids linked together by **peptide** bonds. If there are more than fifty amino acids the molecule is termed a **protein**. Polypeptides are important molecules in the body, e.g. acting as **enzymes**.

Poly(phenylethene) (polystyrene) This is a **polymer** based on the **alkene** phenylethene. It is an addition polymer and is used widely in the form of sheets mainly, in an *expanded* form. Here **air** is blown into the polymer and the result is a white granular solid which has excellent insullation properties. It is used for cups, ceiling tiles and other items which need to be poor **conductors**.

Phenylethene C_8H_8

Poly(propene) This **polymer** is similar to **poly(ethene)** in that it is an addition polymer produced from an **alkene**. The alkene is propene (C_3H_6) which is obtained from petroleum. The polymer is versatile and useful. It is strong and hard wearing and finds use as carpet fibres and material for laboratory flasks and beakers. It is very unreactive. It is also known as polypropylene.

propene C_3H_6

poly(propene)

Polysaccharides These are **polymers** made up of **sugar** molecules such as **glucose**. There are three common naturally occurring examples:

starch
cellulose
glycogen
} These three are all made up of glucose molecules in chains.

Polysaccharides can be broken down by **hydrolysis** into simple sugars such as glucose or into **disaccharides** such as **maltose**. Hydrolysis is brought about by **enzymes** in plants and animals but can also be done by the use of acids.

Polysaccharides are valuable foods, e.g. starch in the form of potatoes or rice; cellulose is found in all plants and glycogen is the form in which excess **carbohydrate** is stored in animals.

Poly(tetrafluoroethene) (PTFE) Polytetrafluoroethene. This **polymer** is *very* inert. It is the non-stick surface for saucepans, e.g. *Teflon, Fluon*. It is also used for bearings because it has a low coefficient of friction.

Polythene See **Poly(ethene)**. The older term polythene is still used in industry and shops.

Potassium Potassium is a **group** I metal. It is a grey, very soft element which is very reactive. It is a powerful **reducing agent**, giving rise to the potassium ion (K^+):

$$2K(s) + 2H_2O(l) \rightarrow 2KOH(aq) + H_2(g)$$

It is stored under oil because of its reactivity towards air and water.

Potassium is obtained by the **electrolysis** of molten potassium chloride (KCl).

Potassium compounds These tend to be colourless, crystalline compounds which are very soluble in water. Potassium **salts** play an important role in the body.

Potassium iodide KI	The solution is a useful solvent for **iodine** forming $KI_3(aq)$. This is used to test for **starch**.
Potassium nitrate KNO_3	*Saltpetre*. This is used in food preservation and in explosives.
Potassium manganate(VII) $KMnO_4$	This is a vivid purple crystalline compound which is a good **oxidizing agent**.

ppt   This is an abbreviation for **precipitate**.

Precipitate   A precipitate is an insoluble subst-ance which is formed in a reaction:

$$Pb(NO_3)_2(aq) + 2NaCl(aq) \rightarrow$$
$$PbCl_2(s) + NaNO_3(aq)$$
$$BaCl_2(aq) + Na_2SO_4(aq) \rightarrow$$
$$BaSO_4(s) + 2NaCl(aq)$$

In these examples, lead(II) chloride and barium sulphate are the precipitates.

Pressure   The pressure on an object is a measure of the force pressing onto its surface. The **SI unit** of pressure is the **pascal**. Other units used are **atmos-pheres** and **mmHg**.

The pressure that a gas exerts upon its container is caused by molecules striking the container's walls.

The pressure of a gas depends upon its volume and its temperature. See **Boyle's law**, **Charles law** and **Gas laws**.

Proof This is a measure of how much **ethanol** there is in a solution of ethanol in water. Spirits such as whisky are normally sold as *70° proof*. This is a 40% solution of ethanol.

Propane Propane is an **alkane** which is obtained from **petroleum**. It is chiefly used as a fuel:

$$C_3H_8(g) + 5O_2(g) \rightarrow 3CO_2(g) + 4H_2O(g)$$

Although a gas at **room temperature**, it is easily liquified and it is sold as bottled gas, e.g *Calor Gas* or *Camping Gaz* where it is mixed with **butane**.

C_3H_8

Properties The properties of a substance are those characteristic ways in which it behaves (reacts) that make it what it is, and make it different from other substances.

It is usual to classify properties as *physical* or *chemical*. Chemical properties are concerned with the substances reactions.

Physical properties include:	Chemical properties include: whether the substance
Colour	
Density	Is a metal or non-metal
Physical state	Gives acidic or basic oxides
Boiling point	Has more than one valency
Melting point	Reacts with acids
Crystal form	Is an oxidizing or reducing
Solubility	agent.
Hardness.	

Proteins These are large, **organic** compounds made up of chains of **amino acids**. The amino acids are linked by **peptide** bonds. They are widespread in animal bodies: skin, hair, nail, wool, gristle muscle.

Proteins form a vital part of our diet. Good sources are: meat, fish, eggs, milk, cheese, beans, bread. The **enzymes** which play so important a role in living organisms are made up of protein molecules. Animals obtain the food which they need in order to make proteins in their bodies by eating other animals or plants. Plants *make* proteins from the nitrogen containing compounds in the soil. See **Fertilizer**.

Proton A proton is a positively charged subatomic particle found in the **nucleus** of the **atom**. The number of protons in an atom is the same as the **atomic number**. **Isotopes** of any element *always* contain the same number of protons.

PTFE See **Poly(tetrafluoroethene)**.

Pure A substance is pure if it contains only one **element** or **compound**, e.g. sodium chloride crystals can be obtained in a pure state, but table salt bought in a packet is impure because it is mixed with other substances.

PVC (polyvinylchloride) See **Poly(chloroethene)**.

Pyrolysis This is the decomposition of a substance by the action of heat:

$$CaCO_3(s) \rightarrow CaO(s) + CO_2(g)$$

Qualitative Qualitative statements or analyses are concerned with composition, *not* with amounts: e.g.

Water is a compound of hydrogen and oxygen.

Quantitative Quantitative statements or analyses are concerned with amounts: e.g.

Water consists of two atoms of hydrogen and one of oxygen.

Radioactivity This is the breaking up of unstable **isotopes** with the emission of radiation or particles.

Some radioactivity occurs naturally, i.e. isotopes found within rocks. Granite usually has quite a large concentration. The types of disintegration which give rise to radioactive emissions are shown:

Electrons (beta particles) can be emitted.
Alpha particles can be emitted.
Gamma rays can be emitted.

See **Nuclear reactions** and **Half-life**.

Rate of reaction The rate (or speed) at which a reaction occurs depends on several factors:

Temperature: The higher the temperature, the faster the reaction is because the particles are moving with greater energies.

Particle size: The smaller the particles involved, the greater is the surface area where the reaction can take place and the faster is the reaction.

Concentration: The more concentrated a solution (or the higher the **pressure** of a gas) the faster a reaction will occur. This is because there are more particles in a given volume to react.

Catalysts: These change the rate by providing an alternative reaction pathway along which the reaction can occur.

Rayon This was one of the first man-made fibres. It is made from **cellulose** (from wood). Wood contains cellulose fibres but they are too short to spin into yarn. Cellulose is dissolved in sodium hydroxide/carbon disulphide solution and squirted into dilute **sulphuric acid**. Here the cellulose is reformed as long threads which are ready for weaving. Rayon is used to make clothes and curtaining.

Reaction When a chemical reaction occurs a new substance is formed. The new substance will have different properties from the reactants.

Reactions can be accompanied by:

- heat
- light
- sound
- colour changes

Reactivity series This is a series of elements arranged in order of their chemical reactivity.

Such a series is shown here:

Potassium
Magnesium
Calcium
Aluminium
Zinc
Iron
Lead
Copper
Silver

The series shown here is a short one containing only a few elements but the same general conclusions apply:

Oxides will be *reduced* by **elements** above them but not the other way round:

$$Mg(s) + CuO(s) \rightarrow MgO(s) + Cu(s)$$
$$Cu(s) + MgO(s) - \text{no reaction}$$

Metal ions in solution will be *reduced* by elements above them but not the other way round:

$$Zn(s) + 2Ag^+(aq) \rightarrow Zn^{2+}(aq) + 2Ag(s)$$
$$Ag(s) + Zn^{2+}(aq) - \text{no reaction}$$

The higher you go up the series, the more vigorous is the reaction with **water** or **acids**.

Potassium reacts violently with cold water, magnesium reacts *very* slowly with cold water but vigorously with steam. Copper reacts with neither.

See **Electrochemical series**. There is a series which involves non-metals.

Recrystallization This is a process for the purification of impure crystalline substances. They are dissolved in a **solvent**, filtered, and then allowed to crystallize. In this way impurities are removed.

Redox A reaction involving both *red*uction and *ox*idation. Any reaction which involves an **oxidation** must also involve a **reduction**. The word *redox* reminds us of this.

Examples of redox reactions:

See **Oxidizing agent, Reducing agent**.

Reducing agent   A reducing agent brings about the **reduction** of a substance. In the reaction it is always oxidized. Common reducing agents:

Hydrogen	H_2
Carbon	C
Carbon monoxide	CO
Sulphur dioxide	SO_2
Hydrogen sulphide	H_2S

Reducing agents at work
$Fe_2O_3(s) + 3CO(g) \rightarrow 2Fe(l) + 3CO_2(g)$
$CuO(s) + H_2(g) \rightarrow Cu(s) + H_2O(g)$
$ZnO(s) + C(s) \rightarrow Zn(s) + CO(g)$

Reduction A substance undergoes reduction if it:

loses **oxygen**	$PbO(s) + C(s) \rightarrow Pb(s) + CO$
gains **hydrogen**	$Cl_2(g) + H_2(g) \rightarrow 2HCl(g)$
gains **electrons**	$Na^+(l) + e^- \rightarrow Na(l)$

See **Oxidation**, **Reducing agent** and **Redox**.

Refining This involves either the removal of impurities from a substance or the removal of components from a mixture.

1) *Metals.* When a metal has been extracted from its ores it is often not in a **pure** state. For some uses this might be acceptable, e.g. **steel** contains impurities. Other uses demand a purer product and then the metal is refined until the right purity is achieved.

2) *Petroleum.* Petroleum is a complex **mixture** of **hydrocarbons**. To make it more useful it is first of all split up into separate *parts*. These *parts* are known as *fractions* and the separation is done by **fractional distillation**. The fractions are also mixtures but are not so complicated. They contain compounds whose **boiling points** are in a certain range.

Conditions vary from refinery to refinery and depend on what products are required. The major product might be **petrol** or it might be petro-chemical feedstock which is then turned into alkenes, etc. for the production of polymers.

See **Cracking** and **Reforming**.

Reforming This is an important process which is used in petroleum refineries and chemical plants. Molecules are taken and altered into more useful products. There is no change in the size of the molecule. One example is the conversion of unbranched alkanes into branched ones, these giving **petrol** a higher **octane rating**:

octane   2-methylheptane

Another example is the production of **aromatic** compounds from **alkenes**:

$$C_7H_{14} \rightarrow \quad C_7H_8 \quad + \quad 3H_2$$
heptene   **methylbenzene** hydrogen

Relative atomic mass (R.A.M.) See A_r.

Relative molecular mass (R.M.M.) See M_r.

Residue This is what remains after any chemical process is complete. The material remaining in a filter paper is called the residue.

Respiration This is the process by which **energy** is obtained by plants and animals.

In animals, food is eaten and broken down in the body. **Oxygen** is breathed in by taking **air** into the lungs. The oxygen is carried round the body in the blood. Reactions occur between the oxygen and the broken-down food and heat is released. **Carbon dioxide** is also produced and this is breathed out. The waste products from the food are excreted from the body. The energy released is used for heat, movement, growth and all other body functions.

energy + waste products

Reversible reactions A reversible reaction is one that can go in either direction, depending on the conditions that exist.

Example:

$$Fe_2O_3(s) + 3H_2(g) \rightleftharpoons 2Fe(s) + 3H_2O(g)$$

Steam can react with hot iron or hydrogen can reduce iron(III) oxide. In both cases, an **equilibrium** will exist with all four substances present unless the products are removed.

Rock salt Impure crystalline deposits of **sodium chloride** are known as rock salt. The solid is purified before use.

Room temperature This means a temperature of between 15°C and 25°C. It is *not* a fixed temperature, but is a range.

Rubber This is a natural **polymer**. It is a **hydrocarbon** and its structure is shown here:

Today most 'rubber' is man-made from butadiene:
$$(CH_2 = CH — CH = CH_2)$$
by **polymerization**. Rubber is elastic and a good insulator. Its uses are widespread from tyres, gloves to waterproofing.

Salt This is a **compound** which is formed when the **hydrogen** of an **acid** is totally or partially replaced by a **metal**, e.g:

$$Zn(s) + HCl(aq) \rightarrow ZnCl_2(aq) + H_2(g)$$

When an acid reacts with a base the products are a salt and **water** only:

$$NaOH(aq) + HNO_3(aq) \rightarrow NaNO_3(aq) + H_2O(l)$$

Salts can also be made by direct combination of two elements:

$$2Na(s) + Cl_2(g) \rightarrow 2NaCl(s)$$

With a **dibasic acid**, if only one hydrogen atom is replaced, the result is an **acid-salt**.

The name of a salt gives a clue as to how it can be formed, e.g:

Magnesium sulphate from magnesium and sulphuric acid:

$$Mg(s) + H_2SO_4(aq) \rightarrow MgSO_4(aq) + H_2(g)$$

Saponification This is the process of the **hydrolysis** of an **ester** when *alkaline* conditions are used. The breakdown of natural fats to produce soap is an example of saponification. Here sodium hydroxide solution is used to produce an alcohol and the sodium **salt** of the **carboxylic acid (soap)**. See **Detergent**.

Saturated compound An **organic** compound containing *only* **single bonds**. Compounds which contain double or triple bonds are said to be **unsaturated**.

Examples

ethane   butan-l-ol

Saturated solution A **solution** which contains the maximum amount of **solute** *at a given*

209

temperature in the presence of excess **solute** is said to be saturated. The amount of solute needed to form a saturated solution depends on the **temperature**. The only way of putting more solute into a saturated solution is by changing the temperature. See **Super-saturated solution**.

Second (s) The **SI unit** of time.

Semi-conductor This is an electrical **conductor** with unusual **properties**. As the temperature increases or the substance contains greater amounts of impurity the resistance of the material *decreases*. In practice, when crystals of semi-conductors are grown, controlled amounts of impurity are added to obtain *exactly* the properties which are required.

Substances which have semi-conduction properties can be **elements** or **compounds** but usually involve **metalloids**.

Examples:

> gallium, germanium, arsenic.

Separation of mixtures Mixtures can be separated if the substances in the mixture have different physical **properties**. Which technique is used depends on the mixture.

Techniques include:
- filtration
- fractional distillation
- chromatography
- immiscibility

210

Shell See **Electronic configuration**.

SI units This is an international system of **units** based on the metric system of measurement. There are seven basic units (shown below) and the system is used in scientific and technological work throughout the world.

Quantity	Unit	Symbol
• length	metre	m
• mass	kilogram	kg
• time	second	s
• current	ampère	A
• temperature	kelvin	K
• amount of substance	mole	mol
• luminous intensity	candela	cd

Silca gel This is a form of silicon dioxide (SiO_2) which can adsorb up to 40% of its own mass of water. It is used as a **dehydrating agent** in **desiccators** and is also used in packaging to maintain goods in a dry atmosphere. It is possible to add a cobalt(II) salt to the gel as this indicates the presence of water, e.g:

red	blue
←	→
wet	dry
hydrated	anhydrous
cobalt(II)	cobalt(II)
chloride	chloride
$CoCl_2.2H_2O$	$CoCl_2$

Silicon   Silicon is a **group (IV) metalloid**. It is a very abundant **element** being a part of the chemical composition of many rocks.

It is a **semi-conductor** and is the heart of micro-electronic technology – the silicon chip.

Silver   Silver is a **transition metal**. It was known to early Man. It is used for jewellery and decorative purposes and has been used for coinage. It is often alloyed with **copper** to give strength.

Silver is low in the **reactivity series**. Its oxide is unstable:

$$Ag_2O(s) \xrightarrow{\text{heat}} 2Ag(s) + O_2(g)$$

Its **halides** are also unstable. This property is used in photography:

$$2AgBr(s) \xrightarrow{\text{light}} 2Ag(s) + Br_2(g)$$

Single bond This is a **covalent bond** which is made up of two **electrons**. Examples are found in all **organic** compounds and in compounds formed from non-metals, e.g:

methane   ammonia   hydrogen chloride

Slaked lime $(Ca(OH)_2)$ This is **calcium hydroxide**. It can be produced by the *slaking* of lime with water:

$$CaO(s) + H_2O(l) \rightarrow Ca(OH)_2(s)$$

A solution of slaked lime in water is known as **lime-water**. Slaked lime is used in agriculture and in making mortar. See **Calcium compounds**.

Soap Soap is a cleaning agent which is made by the action of an **alkali**, e.g. sodium hydroxide solution, on naturally occurring **esters**.

$$H-\underset{\underset{H}{|}}{\overset{\overset{H}{|}}{C}}-\underset{\underset{H}{|}}{\overset{\overset{H}{|}}{C}}-\underset{\underset{H}{|}}{\overset{\overset{H}{|}}{C}}-\underset{\underset{H}{|}}{\overset{\overset{H}{|}}{C}}-\underset{\underset{H}{|}}{\overset{\overset{H}{|}}{C}}-\underset{\underset{H}{|}}{\overset{\overset{H}{|}}{C}}-\underset{\underset{H}{|}}{\overset{\overset{H}{|}}{C}}-\underset{\underset{H}{|}}{\overset{\overset{H}{|}}{C}}-\underset{\underset{H}{|}}{\overset{\overset{H}{|}}{C}}-\underset{\underset{H}{|}}{\overset{\overset{H}{|}}{C}}-\underset{\underset{H}{|}}{\overset{\overset{H}{|}}{C}}-\underset{\underset{H}{|}}{\overset{\overset{H}{|}}{C}}-\underset{\underset{H}{|}}{\overset{\overset{H}{|}}{C}}-\underset{\underset{H}{|}}{\overset{\overset{H}{|}}{C}}-\underset{\underset{H}{|}}{\overset{\overset{H}{|}}{C}}-C\overset{\displaystyle O}{\underset{\displaystyle O^{\ominus}Na^{\oplus}}{\diagdown}}$$

An example of a soapy detergent molecule.

The **hydrocarbon** end of the molecule is **hydrophobic**. The carboxylate end of the molecule is **hydrophilic**.

See **Detergent**.

Sodium   Sodium is a soft, grey **metal** which is in **group** I of the **periodic table**. It is very reactive towards water and non-metals, e.g:

$$Na(s) + 2H_2O(l) \rightarrow NaOH(aq) + H_2(g)$$
$$2Na(s) + Cl_2(g) \rightarrow 2NaCl(s)$$

Sodium is extracted from molten **sodium chloride** by **electrolysis**:

$$2NaCl(l) \rightarrow 2Na(l) + Cl_2(g)$$

The sodium chloride is obtained by mining **rocksalt**, e.g. in Cheshire. Sodium metal is used as a coolant in fast-breeder nuclear reactors.

Sodium is used in the manufacture of the petrol additive tetraethyl lead ($Pb(C_2H_5)_4$) which is used to raise the **octane rating** of fuels. It is also used to extract metals, e.g. titanium:

$$TiCl_4(l) + 4Na(l) \rightarrow Ti(s) + 4NaCl(l)$$

Sodium ions (Na^+) are an important constituent

of the fluids in animal tissues. This is why sodium chloride (common salt) is added to food during the cooking process.

2.8.1

Sodium Compounds

| Sodium carbonate Na_2CO_3 | This compound is made in the **solvay process** and is an important chemical. One of its many uses is in the manufacture of **glass**. Sodium carbonate **hydrate** shows **efflorescence**: $Na_2CO_3.10H_2O(s) \rightarrow Na_2CO_3.H_2O(s) + 9H_2O(g)$ Unlike most carbonates, it is soluble in water. It can be decomposed by heat. Its aqueous solution is *alkaline*. |

Sodium chloride NaCl	Known as *common salt* this compound is obtained from **rock-salt** and is used to make **sodium hydroxide, sodium metal** and **chlorine**. It is used as a flavouring for foods. See **Solvay process**.
Sodium hydrogen-carbonate NaHCO$_3$	This is used in **baking powder**. It is decomposed by the action of heat or **acids**.

$$\text{heat}$$
$$2NaHCO_3 \rightarrow Na_2CO_3(s) + CO_2(g) + H_2O$$
$$NaHCO_3(s) + HCl(aq) \rightarrow NaCl(aq) + H_2O(l) + CO_2(g)$$

Sodium hydrogencarbonate is also used in fire extinguishers and in anti-indigestion powders.

Sodium hydroxide NaOH	This is produced by the electrolysis of **brine** and has wide uses in industry, e.g. making soap and paper. It is a **caustic alkali**. Solutions have pH > 10.
Sodium nitrate NaNO$_3$	This is used as a fertilizer and in the preservation of meat.
Sodium sulphate Na$_2$SO$_4$	The hydrated salt (Na$_2$SO$_4$.10H$_2$O) is known as Glauber's Salt. The sulphate is used in paper manufacture.

| Sodium thiosulphate $Na_2S_2O_3$ | This compound is used in the photographic process. It is used to *fix* the negative and is often known as *hypo*. It reacts with unreacted silver bromide and therefore makes sure that no further reaction with light occurs. |

Softening See the **Hardness** of water.

Solid A solid is a substance whose **atoms** or **molecules** are in fixed positions and do not have the freedom of movement found in a **liquid** or a **gas**. Atoms and molecules are held in a **lattice** by **bonds**. It is only when these bonds are broken that the atoms and molecules are able to move and the solid *melts*.

As **heat energy** is put into the solid lattice the atoms or molecules acquire enough energy to break the bonds. Although the atoms or molecules within the lattice do not move from place to place they do vibrate.

217

Soluble A substance is soluble if it will **dissolve** in a **solvent**. The extent to which any **solute** dissolves depends on the solvent and the temperature. At any given temperature there is a maximum amount of solute which can dissolve. This is a **saturated solution**. See **Solubility**.

Solubility The solubility of a **solute** in a **solvent** is the extent to which it *dissolves*. An **ionic** substance will have a higher solubility in a **polar** solvent than a covalently bonded solute, e.g. **copper** (II) **sulphate** is much more soluble in **water** than is **methane**. The **units** used are **moles** of solute in 100 g of solvent at *a stated temperature*.

Solute A solute is the name given to substances which **dissolve** into a **solvent** to produce a **solution**, e.g. when copper(II) sulphate dissolves in water to produce a solution, copper(II) sulphate is a solute:

$$CuSO_4(s) + H_2O(l) = CuSO_4(aq)$$

Solution When a solid, liquid or gas **dissolves** in a **solvent** to form a solution the particles of the **solute** (i.e. its atoms or molecules) are spread throughout the body of the solvent and are not visible.

Solvay process This process takes cheap raw materials (**brine** and **calcium carbonate** – **limestone**) and turns them into the valuable alkali, **sodium carbonate**:

$$2NaCl(aq) + CaCO_3(s) \rightarrow$$
$$CaCl_2(aq) + Na_2CO_3(aq)$$

Because both sodium carbonate and calcium chloride are soluble in water, the reaction shown cannot be carried out *directly*. It does work, however, by a series of steps:

1) Limestone is heated:
$$CaCO_3(s) \rightarrow CaO(s) + CO_2(g).$$

2) Ammonia is dissolved in brine ($NaCl + NH_3$).

3) Carbon dioxide (produced from the limestone) is passed into the brine/ammonia solution. The following reaction occurs:

$$NaCl(aq) + CO_2(g) + NH_3(g) + H_2O(l) \rightarrow$$
$$NaHCO_3(s) + NH_4Cl(aq)$$

Sodium hydrogencarbonate is produced because of its low solubility. It can be removed from the mixture.

4) Sodium carbonate is produced by heating the hydrogencarbonate:

$$2NaHCO_3(s) \rightarrow Na_2CO_3(s) + H_2O(l) + CO_2(g)$$

5) In the final stage the ammonia is recovered from the ammonium chloride:

$$2NH_4Cl(aq) + CaO(s) \rightarrow$$
$$CaCl_2(aq) + 2NH_3(g) + H_2O(l)$$

The ammonia and carbon dioxide produced in these last two stages are recycled and used again. The only by-product of the reaction is calcium chloride ($CaCl_2$).

Solvent This is a **liquid** in which a **solute dissolves** to form a **solution**. **Water** is the most common solvent. Solvents may be **polar**, e.g. water, or non-polar, e.g. **ether**. Polar solvents dissolve ionic or polar solutes, e.g. **salts**. Non-polar solvents dissolve **covalent** molecules, e.g. **hydrocarbons**.

Spectator ions These are ions which play no part in a reaction. They are only *spectators*. In the example shown below the spectator ions are underlined.

$$\underline{Na^+(aq)} + Cl^-(aq) + Ag^+(aq) + \underline{NO_3^-(aq)} \rightarrow$$
$$\underline{Na^+(aq)} + \underline{NO_3^-(aq)} + AgCl(s)$$

The reaction is: $\boxed{Ag^+(aq) + Cl^-(aq) \rightarrow AgCl(s)}$

Stalactites and stalagmites These are growths which are found in caves. When water

cave roof

stalactite ⟶

stalagmite ⟶

containing **calcium salts** falls into the cave there is a gradual build up of **calcium carbonate** from the roof and the floor as the solution containing the calcium salts decomposes. The stalactites and stalagmites get larger as time passes.

$$Ca(HCO_3)_2(aq) \rightarrow CaCO_3(s) + H_2O(l) + CO_2(g)$$

Starch $((C_6H_{10}O_5)_n)$ Starch is a **carbohydrate polymer**. It is made up of **glucose monomers** and can be broken down by the action of **enzymes** or dilute **acid**. Glucose or **maltose** is formed. Starch is insoluble in water and is a white tasteless powder. It is found in most plants and is used by plants as a food. Animals also use it as food when they eat plants, e.g. potatoes, rice, flour.

In the presence of starch, **iodine** solution turns blue. This can be used as a test for iodine or starch. See **Polysaccharide**.

States of matter Substances can exist in three states of matter. These are **solid, liquid** and **gas**. Substances are changed from one state into another by altering the **temperature**.

Steel Steels are **alloys** which contain **iron** as the main constituent. Depending which other **elements** are present, substances of varying properties can be formed. The two best known steels are *mild steel* which is used for car bodies and household goods such as cookers, freezers, etc. and *stainless steel* which is used in industry and in cooking utensils. Mild steel rusts easily and so has to be protected by **galvanizing**, enamelling or painting. Stainless steel is not corroded by oxygen and so is very useful.

The iron made in the blast furnace contains impurities which make it brittle, e.g. carbon. Iron is made into steel by blowing oxygen through the molten iron and thus oxidizing the impurities. Then, the correct amounts of other elements can be added to produce the steel required, e.g. stainless steel.

STP This stands for *S*tandard *T*emperature and *P*ressure. The conditions it refers to are one **atmosphere** pressure and a **temperature** of 0°C (273 K). When comparing gas volumes it is useful to have *standard conditions* to refer to.

Strengths of acids and bases Acids and **bases** are termed *strong* if they are fully dissociated in solution. Good examples are acids such as hydrochloric, nitric and sulphuric and bases such as sodium or potassium hydroxide.

$$HCl(aq) \rightleftharpoons H^+(aq) + Cl^-(aq)$$
pH range 1–2

$$NaOH(aq) \rightleftharpoons Na^+(aq) + OH^-(aq)$$
pH range 13–14

Weak acids and bases do not fully dissociate in solution. Here the equilibria lie on the *left-hand* side. Good examples are found in the organic acids, e.g. **citric**, tartaric, **ethanoic** and malic; and in bases such as ammonia.

$$CH_3COOH(aq) \rightleftharpoons CH_3COO^-(aq) + H^+(aq)$$
pH range 3–6
$$NH_3(aq) + H_2O(l) \rightleftharpoons NH_4^+(aq) + OH^-(aq)$$
pH range 8–11

Strong acid and base See **Strengths of acids and bases**.

Structural formula A structural formula is one that shows the **bond** between atoms and the position of the atoms with respect to each other. The **molecular formula** shows the number of atoms in the molecule.

ethane
C_2H_4

$$\begin{array}{ccc} & H & H \\ & | & | \\ H-&C-&C-H \\ & | & | \\ & H & H \end{array}$$

ethanol
C_2H_5OH

$$\begin{array}{ccc} & H & H \\ & | & | \\ H-&C-&C-O-H \\ & | & | \\ & H & H \end{array}$$

Sub-atomic particles These are particles which are found in the atom. The important ones are the **proton, neutron** and **electron**.

Sublime If a **solid** changes directly into a **gas** when it is heated it is said to sublime. Common substances which sublime are:

Carbon dioxide CO_2 Iron(II) chloride $FeCl_2$

Substitution reaction This is a reaction where

223

some atoms or groups in a molecule are replaced by others. It is a common reaction of **organic** compounds.

Example:

$$CH_4(g) + Cl_2(g) \rightarrow CH_3Cl(l) + HCl(g)$$

A C–H bond is replaced by a C–Cl bond.

Sucrose $(C_{12}H_{22}O_{11})$ This is a **disaccharide** which is made up of a **fructose** unit and a **glucose** unit. It is the white crystalline **sugar** that is used in the home. It is obtained from sugar beet and cane sugar.

Sulphates These compounds contain the SO_4^{2-} ion (**valency** = 2). They are widespread in nature, e.g. gypsum – $CaSO_4$. They can be produced in the laboratory by the action of **sulphuric acid** on **metals** or **oxides** and **hydroxides**, e.g:

$$Zn(s) + H_2SO_4(aq) \rightarrow ZnSO_4(aq) + H_2(g)$$
$$CuO(s) + H_2SO_4(aq) \rightarrow CuSO_4(aq) + H_2O(l)$$

The test for a sulphate is to add a solution to an acidified (HCl) solution of barium chloride. If a sulphate is present a white **precipitate** is formed:

$$BaCl_2(aq) + SO_4^{2-}(aq) \rightarrow BaSO_4(s) + 2Cl^-(aq)$$

This precipititate is insoluble in dilute acid.

Sulphides These are compounds of sulphur with another **element**. They are produced by reaction with **sulphur** or **hydrogen sulphide**:

$$Fe(s) + S(s) \rightarrow FeS(s)$$
$$CuO(s) + H_2S(g) \rightarrow CuS(s) + H_2O(g)$$

Sulphites These compounds contain the SO_3^{2-} ion and can be thought of as **salts of sulphurous acid**. **Sulphur(IV) oxide** is prepared in the laboratory by the action of **acid** on a sulphite.

Sugar This is the common general term for those sweet compounds which chemically are **monosaccharides** and **disaccharides**. Examples: **glucose, sucrose, fructose, maltose**. See **Carbohydrate**.

Sulphur (S_8) Sulphur is a yellow non-metallic **element** which exists as two allotropic forms. Rhombic sulphur is the stable form below $96°C$ and monoclinic sulphur is stable above that temperature. The figures in the chart refer to the monoclinic **allotrope**.

Sulphur is a **group** (IV) of the **periodic table** and is reactive towards metals and oxygen. It is found uncombined in nature as well as occurring as metal sulphides, e.g. galena (PbS) and pyrite (FeS).

Sulphur containing compounds are also found in **petroleum** and **natural gas**. Sulphur is an important constituent of some drugs, e.g. sulphonamides but its major use is in the manufacture of **sulphuric acid** in the **contact process**. See **Vulcanizing**.

Sulphur compounds

Sulphur(IV) oxide SO_2	This is a colourless gas with a sharp odour. It is produced by burning **sulphur** or sulphur compounds in **air** or **oxygen**. It dissolves in water to form **sulphurous acid** – a weak acid. It is useful as a sterilizing agent and is often added to soft drinks. It is a **reducing agent**. This property is used in the chemical test for the gas. Sulphur(IV) oxide will *reduce* dichromate(VI) ion (orange) to chromium(III) ion (green).
Sulphur(VI) oxide SO_3	This is a **volatile** solid with a sharp odour. It is produced in the **contact process** by the combination of **sulphur dioxide** and **oxygen** over a **catalyst**. The oxide is acidic and when it is added to water, sulphuric acid is formed. $H_2O(l) + SO_3(s) \rightarrow H_2SO_4(aq)$

Sulphuric acid (H_2SO_4) Sulphuric acid is one of the most important chemicals produced. It is a colourless, oily liquid which is a **strong acid** and a vigourous **oxidizing agent**. It is made in the **contact process** and the annual production in the UK is about four million **tonnes**. The chart shows the main uses of the acid and the approximate amounts of the acid which are used.

fertilizers
30% e.g. ammonium sulphate
superphosphate

17% paints

12% natural and man-made
fibres

11% detergents

11% other chemicals e.g.
drugs, explosives

10% miscelaneous uses
e.g. car battery

3% plastics
3% metallurgical uses
2% dye STUFFS
1% petroleum refilling

Sulphuric acid reacts chemically in several ways.
As an acid Dilute sulphuric acid reacts with **metals, bases** and **carbonates** to form sulphates:

$$Mg(s) + H_2SO_4(aq) \rightarrow MgSO_4(aq) + H_2(g)$$
$$CuO(s) + H_2SO_4(aq) \rightarrow CuSO_4(aq) + H_2O(l)$$
$$ZnCO_3(s) + H_2SO_4(aq) \rightarrow$$
$$ZnSO_4(aq) + H_2O(l) + CO_2(g)$$

Concentrated sulphuric acid reacts with chlorides and nitrates to form hydrogen chloride and nitric acid respectively:

$$H_2SO_4(l) + NaCl(s) \rightarrow NaHSO_4(s) + HCl(g)$$
$$H_2SO_4(l) + NaNO_3(s) \rightarrow NaHSO_4(s) + HNO_3(g)$$

As a dehydrating agent The concentrated acid is **hygroscopic**. Concentrated sulphuric acid is sometimes used to dry gases. It can also be used to remove water from substances. Hydrated copper(II) sulphate can be dehydrated:

$$CuSO_4.5H_2O(s) \rightarrow CuSO_4(s) + 5H_2O(l)$$
$$\text{blue crystals} \qquad\qquad \text{white powder}$$

methanoic	carbon
acid	monoxide

If the concentrated acid is added to **sucrose**, the sugar turns black. This is because water is removed from the **carbohydrate** and **carbon** is left behind:

$$C_6H_{12}O_6(s) \rightarrow 6C(s) + 6H_2O(g)$$

The reaction between concentrated sulphuric acid

and water is very **exothermic**. It is important always to add the acid to water and *not* the other way round.

As an oxidizing agent Although copper cannot displace hydrogen from acids, the metal can be oxidized by concentrated sulphuric acid:

$$Cu(s) + 2H_2SO_4(l) \rightarrow CuSO_4(s) + SO_2(g) + 2H_2O(l)$$
anhydrous
copper(II)
sulphate

Sulphurous acid (H_2SO_3) This acid only exists in aqueous solution. It is formed by passing **sulphur dioxide** into water.

It is a **weak acid** and a **reducing agent**. When heated, sulphur dioxide and steam are produced, leaving no residue. Salts of the acid are called sulphites.

Supersaturated solution This is a solution which contains a higher **concentration** of **solute** than a **saturated solution**. It is usually produced by cooling the saturated solution. If the solution is disturbed, e.g. by a mechanical shock or by dust falling into it or by a crystal of the solute being added to it, the excess solute will usually crystallize out.

Suspension When a **solid** is added to a **liquid** and the solid neither dissolves in the liquid nor sinks to the bottom of the vessel, the mixture is referred to as a suspension because the solid is *suspended* in the liquid.

solid particles suspended in the liquid

Symbol Symbols are often used in chemistry to represent the names of substances, e.g. **elements**.

> Fe – Iron
> S – Sulphur
> O – Oxygen

Symbols are also used in representing units.

Synthesis This term means that production of something from smaller or simpler parts, e.g. the synthesis of **ammonia** from hydrogen and nitrogen.

Synthesis gas This is a mixture of **carbon monoxide** and **hydrogen** gas produced by the steam reforming of **natural gas**. It is used to synthesize other compounds.

Synthetic *Man-made*, e.g. man-made fibres such as **nylon**, etc.

Temperature The temperature of a substance is the measure of its **kinetic energy**. The temperature scale used in everyday life is the *Celcius* (or *Centigrade*) scale. In scientific work, the **Kelvin** scale is usually used.

Tempering   This is when a hot **metal** is allowed to cool slowly in air. By doing this, the metal becomes tough and springy.

Terylene   This is a **polyester** which is widely used in the clothing and household furnishing industries. It is often mixed with natural fibres, e.g. Terylene/wool mixtures. Its monomers are a dialcohol and a diacid. The formula of the polymer is:

from the   from the
diacid   dialcohol

Tetrachloromethane   Tetrachloromethane is a dense, colourless liquid which is **immiscible** with water. It is a good solvent for non-**polar** com-

CCl₄

Cl
|
Cl—C—Cl
|
Cl

pounds. It can be made from methane by a **substitution reaction** involving **chlorine**:

$$4Cl_2(g) + CH_4(g) \rightarrow CCl_4(l) + 4HCl(g)$$

The other substituted methane products are also formed: (CH_3Cl, CH_2Cl_2, $CHCl_3$) and the mixture must be separated to obtain **pure** products.

Tetrahedral In a carbon compound such as methane where there are four **single bonds** leading to four atoms, the atoms are arranged tetrahedrally around the carbon atom. The bonds point to the corners of the imaginary cube that the carbon atom is in the centre of. In this way, the atoms are as far away from each other as possible. The angle between the bonds is 109° 28′ – the tetrahedral angle.

CH_4
methane

Thermal This is a word which means 'to do with heat', e.g. thermal decomposition – decomposition by heat. Thermal **energy** – heat energy.

Thermit reaction In this reaction aluminium removes the oxygen from the oxide of a metal below it in the **reactivity series**. Iron(III) oxide is usually used in the laboratory and was used in industry to provide a small source of molten metal. Great heat is liberated in the reaction and the products are iron and aluminium oxide.

$$Fe_2O_3(s) + 2Al(s) \rightarrow Al_2O_3(s) + 2Fe(s)$$

In the reaction in the laboratory a 'fuse' of magnesium is usually used and barium peroxide is used to start the reaction off. Once started, the reaction is very **exothermic**.

magnesium fuse

barium peroxide + powdered magnesium

Al + Fe_2O_3

sand-filled bucket

Thermometer Thermometers measure **temperature** – i.e. how hot something is. Although they come in different forms, the most common are the 'liquid in glass' thermometers.

The liquid – usually **mercury** – is held in a bulb. As the temperature rises, the liquid expands in the bulb and rises up the tube. The tube is marked at intervals with the temperature and this can be read off.

marks on the glass- 'graduations' to show the temperature

small tube into which the liquid expands

reservoir for the liquid 'bulb'

Thermoplastic A thermoplastic **polymer** is one which softens when it is heated and can be moulded and re-moulded into new shapes. Examples: **nylon, poly(chloroethene)**.

Thermosetting A thermosetting **polymer** is one which cannot be softened again once it has been heated. Decomposition occurs if it is heated again. Examples: bakelite, formica.

Tin Tin is a **metal** which is in **group** IV of the **periodic table**. It occurs as two **allotropes** – grey tin and white tin. The grey form is stable at low temperatures and the data given in the diagram refers to the white form.

White tin is a shiny metal which shows normal metallic reactions. It is found naturally as the oxide SnO_2.

Tin is used to make tin plate (**steel** with a very thin layer of tin upon it), **bronze**, solder and **alloys** for bearings. It is also used in the production of **glass**. Glass is floated onto the surface of molten tin. In this way large areas of glass without defects are formed. Tinplate is used to make cans and metal boxes in which food is stored.

Titanium This is a **transition metal** which is

235

common in the earth's crust (0.5%). It forms **alloys** which are very strong while having a low **density**. It is resistant to **corrosion** up to a high temperature. Titanium alloys are used in the aircraft industry. TiO_2 is used as a pigment.

Titration   A titration involves the reaction of two **solutions** and is used in analysis to discover the **concentration** of one of the solutions. One solution is normally placed in a conical flask and the other in a **burette**. The solution from the burette is added slowly until the other solution has been completely used up. This is shown by the use of an **indicator** or

burette containing a solution whose volume is known but whose concentration is not known

clamp

tap

conical flask containing a solution whose volume and concentration are known by means of a **pH** meter or a conductivity meter. The technique is most commonly used for acid-base reactions. See **Volumetric analysis**.

Tonne This is a **unit** of mass. One tonne equals 1000 kilograms. It is 2205lb and so is 35lb lighter than the imperial ton (2240lb). The tonne (sometimes called the *metric ton*) is now widely used.

Transition metals These are the **elements** found in the central section of the **periodic table**. There are three series of them but the most important are those elements from scandium to zinc. Transition metals have similar **properties**. They

- produce coloured **compounds**
- have variable **valency**
- are high **melting** and **boiling point** solids
- are useful as **catalysts**
- have **cations** which are useful as catalysts.

They tend to be important elements for use as metals, e.g. iron, copper, silver, gold, mercury, zinc, platinum, or for use in alloys, e.g. titanium, vanadium, chromium, manganese, cobalt, nickel, tungsten.

Transition temperature This is the **temperature** above which one **allotrope** is stable and below which another is stable. Some examples are shown here.

rhombic sulphur stable below	96°C	monoclinic sulphur stable above
grey tin stable below	19°C	white tin stable above

Trichloromethane This molecule is also known as *chloroform* and has been used in anaesthetic work.

Trichloromethane is a substituted alkane. It is a colourless, volatile liquid which boils at 62°C. See **Tetrachloromethane**.

Triple bond A triple bond contains six **electrons** and is found in **alkynes** and nitrogen-containing compounds.

Examples:

ethyne $H - C \equiv C - H$
nitrogen $N \equiv N$

Ultraviolet radiation (uv) This is invisible

radiation which has a slightly higher frequency (energy) than violet light. It is produced in large amounts by stars, e.g. the sun. The radiation in large amounts is harmful to humans – skin cancers can result. Not much of the uv radiation from the sun reaches the earth. Much of it is filtered out by the **ozone** layer in the upper **atmosphere**.

Units These are quantities or measurements which are used as standards to measure other things.

Measurement	Unit
distance	metre
time	second
mass	kilogramme

See **SI**.

Universal indicator This is a mixture of **indicators**. Because it is a mixture it changes colour several times. It is possible to tell the approximate **pH** of a solution by adding a few drops of universal indicator to it and reading the pH off a chart such as the one shown here. Universal indicator paper strips are also available:

Colour	← red orange yellow green blue purple →
pH	0 1 2 3 4 5 6 7 8 9 10 11 12 13 14

Unsaturated compounds Carbon compounds which possess **double** or **triple bonds** are said to be unsaturated. Some examples are:

aldehydes and ketones

$$\begin{array}{ccc} \diagup C = C \diagdown & -C \equiv C- & -C \equiv N \\ \textbf{alkenes} & \textbf{alkynes} & \text{cyanides} \end{array}$$

They contain more **electrons** in their bond than a normal **single bond**. Because of this they are reactive. They tend to react through **addition reactions** forming new bonds with the electrons they possess, e.g. **alkenes** become **alkanes**.

ethene ⟶ **ethane** (saturated)

Uranium Uranium is a metallic element which has three naturally-occurring **isotopes** (masses 234 – trace, 235 – 0.7%, 238 – 99.3%).

They are all radioactive and eventually decay to give stable isotopes of **lead**.

Uranium is used as a fuel in nuclear power stations. In some types the oxide U_3O_8 is used. In others uranium containing a higher proportion of

U-235 is used (enriched uranium). In the fast-breeder reactor uranium 238 is turned into **plutonium** 239 which can then be used as a fuel.

Vacuum A vacuum is where there are no **atoms** or **molecules** present. It is impossible to obtain a *perfect* vacuum but we talk of a *partial* vacuum when the **pressure** is extremely low.

The pressure is low in space and this is usually thought of as a vacuum.

Valency The valency of an **element** is the number of **electrons** that it needs to form a **compound** or radical. The electrons can be *given to* another element; they can be *taken from* an element; they can be *shared*.

Some elements always have the same valency, e.g. hydrogen = 1, oxygen = 2 (except in peroxides), sodium = 1, magnesium = 2.

Non-metals tend to have more than one valency, e.g. sulphur = 2, 4, 6; nitrogen = 1, 2, 3, 4, 5.

241

Examples:

Sodium has a valency of one. Sodium gives one electron away when it forms the sodium **ion** Na^+.

Oxygen has a valency of two. Oxygen accepts two electrons when it forms the oxide ion O^{2-} or it forms covalent compounds, e.g. SO_2, CO_2, N_2O.

Hydrogen and bromine have valency one. They both give one electron to make the H–Br bond when they form hydrogen bromide.

Transition elements have more than one valency: Iron = 2, 3; cobalt = 2, 3; manganese = 2, 3, 4, 5, 6, 7; copper = 1, 2.

Vanadium Vanadium is a **transition metal**. Its chief use is in the production of vanadium-steel **alloys** which are valuable because of their high tensile strength and hardness. They often include **chromium** or **manganese** in addition to vanadium.

Vanadium(v) oxide V_2O_5 is the **catalyst** used in the **contact process**.

Van der Waals' forces These are weak forces between **atoms**. They are caused by the movement of **electrons** within the atoms. Elements which only have these forces to hold them together, e.g. the **inert gases**, have very *low* **melting** and **boiling points**.

Vapour A vapour is **atoms** or **molecules** in the gaseous state *but* below their critical temperature.

For example, when water evaporates from a saucer when it is left in the house, *water vapour* is formed and not steam.

Vinegar This is a solution which is made by the action of bacteria on wine or cider. It contains about 4% **ethanoic acid**. It is used widely in the food industry for preserving foods.

Vitamin Vitamins are chemicals which are important to the proper working of the body. They tend to be complex **organic** molecules which cannot be made in the body but which must be eaten.

Some examples:
   Vitamin A found in dairy foods and liver.
   Vitamin C found in fresh fruit and vegetables.
Absence of vitamins cause deficiency diseases.

Volatile A substance which is volatile is easily turned into a **vapour**. Such substances either have

boiling points which are near **room temperature**, e.g. **ether** or propanone; or they are solids which **sublime**, e.g. **carbon dioxide**. Liquids which are volatile *and* flammable are very dangerous because of the risk of **explosions**. Care has to be taken with their storage and handling. **Petrol** is such a liquid.

Volume　　The volume of a substance is the amount of space that it occupies. Solids and liquids have fixed volumes but a gas will have the same volume as the container it occupies. The larger the container is, the *lower* is the **pressure**. The volume of a gas can easily be changed by compressing it but it is much more difficult to change the volumes of solids and liquids. *Hydraulic* brakes in a car depend on this property of a liquid. Volumes are measured in cubic centimetres (**cm³**).

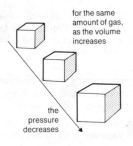

for the same
amount of gas,
as the volume
increases

the
pressure
decreases

Volumetric analysis This is a method of **quantitative** analysis which uses accurately measured volumes of solutions. See **Titration, Burette** and **Pipette**.

Vulcanization In this process **sulphur** is added to **rubber** to make it harder.

Washing soda This name is given to **hydrated** sodium carbonate ($Na_2CO_3.10H_2O$). The name comes from the use of the salt to soften water which was to be used for washing.

Water (H_2O) Water is an oxide of **hydrogen**:

$$2H_2(g) + O_2(g) \rightarrow 2H_2O(l)$$

It is one of the most common compounds on earth. It is the best known **solvent** and is needed by all living things.

Water is a colourless liquid and some of its more important properties are shown here:

Freezing point 0°C
Boiling point 100°C
Density 1.0 g/cm³
(water has a maximum density at 4°C)

It does not conduct electricity although it can be electrolysed if small amounts of acid (H_2SO_4) or alkali (NaOH) are added. The products are hydrogen and oxygen:

$$2H_2O(l) \rightarrow 2H_2(g) + O_2(g)$$

Two tests for the presence of water are by changing the colour of **anhydrous salts**:

$$CuSO_4(s) + 5H_2O \rightarrow CuSO_4.5H_2O(s)$$
white   blue

$$CoCl_2(s) + 6H_2O \rightarrow CoCl_2.6H_2O(s)$$
blue   red

These tests show that water is present and not that the water is **pure**. To show purity, the **boiling point** of the liquid could be taken.

Water is found in the **atmosphere**, in lakes, rivers, glaciers and the oceans. It is found in rocks and in living creatures. Water is continually moving from place to place on the earth – see the **water cycle**.

The most important chemical property of water is its use as a **solvent**. Water has **polar bonds** and so can dissolve ionic solids such as sodium chloride (NaCl) as well as polar solids such as **glucose** ($C_6H_{12}O_6$).

The water which we drink is never pure. It always contains small amounts of gas (e.g. oxygen and carbon dioxide) and depending on the **hardness** of the water, solids are dissolved in it too.

Water cycle   There is a continual movement of water around the earth, both in the oceans and in the **air**. Water falls to earth as rain, snow, hail, sleet, and freezes out of the air as frost and ice. It falls onto the oceans and these act as a vast store of water. It falls

sea

evaporation

clouds

combustion
respiration

transpiration

town

river

rainfall

streams
and rivers

lake

snow

ice cap

on land and here it enters the earth and eventually flows into lakes and rivers and then flows into the oceans. From these large areas of water, evaporation occurs and water re-enters the **atmosphere**. In this way the cycle continues. Plants and animals take water out of the ground for their own use and it can re-enter the atmosphere (*transpiration, excretion* and **respiration**) or the earth. Large quantities of water are stored as glaciers and the polar ice caps. The cycle is shown diagramatically:

Water of crystallization   This is water which is chemically bonded within **crystals**.

Examples are:

$$CuSO_4.5H_2O \quad Na_2CO_3.10H_2O \quad CaSO_4.2H_2O$$

Water can be removed by heating, leaving the **anhydrous** salt. This can happen because the chemical bonds are weak. Some **salts** which include water of crystallization lose some of the chemically bound water because of **efflorescence**.

Weak acids and bases   See Strengths of acids and bases.

X-rays   This is *electromagnetic radiation* which has high **energy**. It is produced by firing **electrons** at **metals**. It is a very penetrating radiation and will easily pass through flesh but is stopped by bone and other dense substances such as metals. X-rays are useful in medicine in checking for broken bones and also for objects which have been swallowed.

Xenon Xenon is one of the few **inert gases** which have been found to form compounds. XeF_4 has been produced. This is possible because the xenon has massive atoms which can form **van der Waals'** bonds with other atoms such as **fluorine**. The element is used to fill light bulbs and fluorescent tubes.

Yeast Yeasts are microscopic organisms which are very useful to man. They are used in baking and brewing. In brewing, they are used to convert **sugars** into **ethanol**:

$$C_6H_{12}O_6(aq) \xrightarrow{\text{yeast}} 2C_2H_5OH(aq) + 2CO_2(g)$$

glucose   **ethanol**   **carbon dioxide**

The results are alcoholic drinks such as beer and wine. The carbon dioxide is either collected and sold or it is allowed to escape into the **atmosphere**.

In baking, it is the carbon dioxide which is useful. The dough is mixed and then the sugar and yeast react in the dough to produce the gas. The gas makes the dough expand (rise) and the mixture becomes much lighter.

Yeasts contain **enzymes** and it is these substances which act on the sugars. See **Zymase**.

Yield of a reaction   Many chemical reactions do not produce as much product as would be expected from looking at the chemical **equations**. It is usual to express the amount of product as a percentage of what it is theoretically possible to produce:

$$\frac{\text{Yield of}}{\text{reaction}} = \frac{\text{amount of product produced}}{\text{maximum amount of product that it is possible to produce}}$$

Example:
$$CuO(s) \quad + \; H_2(g) \quad \rightarrow Cu(s) \; + \; H_2O$$
copper(II)   hydrogen   copper   water
oxide
80 g   → 64 g (maximum yield)

If 80 g of oxide actually produce 56 g of copper metal the percentage yield of the reaction is $(^{56}/_{64} \times 100)\% = 87.5\%$.

Z   This is the symbol which is given to represent the **atomic number** of an **element**, i.e. the number of **protons** in the **atom**.

Zinc   Zinc is a **transition metal** which is found in the *first* transition series in the periodic table.

It is a dense grey metal which is reactive towards **acids** but which does not react with cold water. The metal is used in the **alloy, brass** and also in the protection of **steel** by **galvanizing**. Zinc is extracted from the ore *galena* (ZnS).

Zinc compounds

Zinc oxide ZnO	This has use in medicine – zinc oxide cream. It is used as a protection against skin irritations, e.g. nappy rash. It is also used in paints. Zinc oxide and hydroxide are **amphoteric**.

Zymase   This is the **enzyme** which is present in yeasts which is responsible for the formation of **ethanol** and carbon dioxide from **sugars**.